T0291660

CAMBRIDGE COMPARATIVE PHYSIOLOGY

GENERAL EDITORS:

J. BARCROFT, C.B.E., M.A.
Fellow of King's College and Professor of
Physiology in the University of Cambridge
and
J. T. SAUNDERS, M.A.
Fellow of Christ's College and Demonstrator
of Animal Morphology in the University of
Cambridge

THE
COMPARATIVE PHYSIOLOGY
OF INTERNAL SECRETION

THE
COMPARATIVE PHYSIOLOGY
OF INTERNAL SECRETION

BY

LANCELOT T. HOGBEN
M.A. (Cantab.), D.Sc. (Lond.)
PROFESSOR OF ZOOLOGY IN THE
UNIVERSITY OF CAPETOWN

CAMBRIDGE
AT THE UNIVERSITY PRESS
1927

CAMBRIDGE
UNIVERSITY PRESS

University Printing House, Cambridge CB2 8BS, United Kingdom

Cambridge University Press is part of the University of Cambridge.

It furthers the University's mission by disseminating knowledge in the pursuit of education, learning and research at the highest international levels of excellence.

www.cambridge.org
Information on this title: www.cambridge.org/9781107502277

© Cambridge University Press 1927

First published 1927
First paperback edition 2015

A catalogue record for this publication is available from the British Library

ISBN 978-1-107-50227-7 Paperback

TO
E. W. M.

CONTENTS

Note—Fig. 27 *a.* (Carotid blood pressure.) *faces page* 88

ERRATUM

Page 14, line 21, *for* **26** *read* **24**.

Chapter I

CHEMICAL CO-ORDINATION

§ I

In practically all cellular animals other than sponges visible manifestations of activity involve a receptive surface on which the stimulus operates, a structure specialised for the performance of the appropriate response, and, intervening between these, a mechanism of co-ordination. Co-ordination in animals is of two kinds. Cellular animals in general possess specialised tissues in continuity with both receptor and effector units: this constitutes the nervous system. A large number of animals also possess a special arrangement, the circulatory system, for keeping in motion the fluids which bathe the tissues: the circulatory system provides an avenue through which not only general metabolic changes but active responses of the organism can be regulated by the distribution of specific excitants which are capable of producing local responses of one kind or another. Chemical or endocrine co-ordination is an important aspect of the regulatory processes of the body. Though in some respects more accessible to experimental investigation than the analysis of the nervous function, it is a subject of more recent growth, and one in which our knowledge is at present restricted almost exclusively to vertebrate species. This is partly because research in this field has been motivated in the main by clinical objectives. But there is, it is hoped, sufficient information drawn from materials other than man's nearest of kin to justify a survey from the standpoint of the comparative physiologist.

In embarking upon a study of the rôle of internal secretion in co-ordinating the activities of the organism, it is important to define the precise sense in which the term *internal secretion* will be employed in these pages, since it has been used by some authors in such an indefinite manner as to cover the whole field of intermediate metabolism, a subject which is dealt with in a separate volume of this series. It is the intention of the writer to restrict the term for the purpose of the present discussion to the production of substances which are liberated into the blood stream by the specific activity of a particular structure (endocrine organ), and, when set free in the circulation, are capable of evoking responses in tissues remotely situated from their point of origin. So defined, the term *internal secretion* is in effect synonymous with *hormone*. The latter is sometimes wrongly employed to signify any physiologically active tissue constituent (e.g. the pressor component in pituitary extracts) whether or not it has been proved to have a functional significance or to be liberated into the blood stream. For excitant substances present in animal tissues it is preferable to employ the more general term due to Schafer, *autacoid*. The preparation and investigation of the properties of autacoids is the subject matter of several comprehensive and authoritative works; and information of this kind is of the utmost importance for the furtherance of our knowledge of internal secretion. It is also valuable, now that the subject has progressed so far, to take a retrospective view, and attempt to formulate the extent to which the phenomenon of internal secretion can rightly be regarded as an effective agent in the co-ordination of vital activities. Such treatment must neces- sarily be orientated with reference to the various modes of activity which the organism displays; and inevitably much matter of a kind which is dealt with in standard works on endocrinology will be omitted from the present account.

Since we are here concerned with the functional significance of internal secretion as part of the mechanism of co-ordination, it will be as well to devote some space at the outset to a critical examina- tion of the type of evidence upon which it is customary to rely for proof of endocrine function. Suggestive indications of the possi- bility than an organ is an organ of internal secretion in the sense defined above is furnished by a study of the pharmacodynamic properties of tissue extracts and the characteristic disturbances which accompany diseased conditions or operative removal of glandular structures. Investigation of the first type owes its origin to the researches of Oliver and Schafer (1894–5). Information of the second kind dates from before the dawn of medicine to those who introduced the practice of castration—probably in the first place as a religious observance. Though isolated information of this kind is actually the only basis for the customary description of several familiar organs as ductless glands, neither of these criteria can of itself suffice to constitute a rigid proof of endocrine activity.

Thus the fact that extracts of fresh pituitary glands from all classes of land Vertebrates, if suitably prepared, have the specific effect of producing rise of blood pressure in the Mammal and fall of blood pressure in the Bird, is a very good reason for exploring the possible relation of the pituitary to vasomotor regulation. It is not sufficient to justify the inference that such a relation does in fact exist. In pursuing enquiry into the endocrine or supposedly endocrine function of so-called ductless glands the physiologist has no more justification for attributing a teleological significance to every chemical entity in the organism than neo-Darwinian naturalists had for ascribing utility to every member of the body. Apart from this purely formal issue, there are certain technical difficulties to be recognised before sorting out data with reference to autacoid substances. In all tissues after death there are formed by autolysis and putrefaction substances which have pronounced physiological effects. One may say that all commercial preparations contain traces of histamine-like and choline-like compounds. Consequently statements regarding the properties of extracts prepared from an organ should be accepted with caution, unless based upon absolutely fresh material and quantitative comparison with the activity of similarly prepared extracts from other tissues.

Again, experimental removal of organs and the effects of disease can provide valuable indications of their functional importance; but cannot in the absence of collateral evidence suffice to prove that they liberate hormones into the circulation. To infer endocrine activity from this source alone is hardly less unwarranted than to deduce from the manifest consequences of decapitation the conclusion that the head secretes a hormone that maintains the rhythmical contraction of the heart. Nevertheless in one instance this line of argument has been almost universally adopted. With the notable exception of the late Geoffrey Smith nearly every investigator in the field of sex differentiation has referred to internal secretions of the ovary or testis, as if the existence of such secretion were an established fact. What is clearly established is the experience that removal of the ovary or testis prevents the appearance of certain sexual characteristics, as illustrated by the assumption of male plumage, spurs, etc., after spaying in the hen, and inhibition of the growth of the antlers in the castrated male deer, etc. It is also clear that in the Mammal the interstitial tissues are pre-eminently involved. All that can be legitimately inferred from these facts is the conclusion that in some way or other the presence of an ovary or testis determines the way in which sex differentiation proceeds. The development of secondary sex characters—except in the special case where they depend upon sex-linked factors—is the result of differences of metabolism acting upon material of similar genetical constitution. Such differences arise in virtue of the presence of one or other type of gonad. Since we are still quite ignorant as to the nature of the metabolic difference which under-lies one or other type of sex differentiation, no economy of hypo-thesis is effected by assuming that the gonad discharges specific exciting substances into the circulation in preference to the equally plausible alternative that it removes something.

On the basis of his researches on sex transformation in crabs parasitised by *Sacculina*, Geoffrey Smith (1913) pointed out that the blood of the normal female of *Inachus* differs from that of the male in its higher content of certain lipoid constituents ab-sorbed by the ovaries for the formation of the yolk of the egg. These fatty substances form an important part of the food of *Sacculina*, and since normal metabolism involves their production in large quantities as fast as they are used up in yolk formation by the female, the action of the parasite in feminising the male crab would appear to imitate the consequences resulting from the presence of an ovary in stimulating the production of lipoid materials. The parasite absorbs them as fast as they are produced, the only result in the female being the degeneration of the ovary. But in the case of the male the character of the metabolic processes is so transformed that their precursors are produced in quantities comparable with those present in the blood of the normal female, to the somatic organisation of which the crab approximates. It may be doubted whether conclusive proof is brought forward by Geoffrey Smith in support of the proposition that these products of intermediate metabolism are the essential factors in directing somatic differentiation in the direction of the female or male con-dition. But some such interpretation of the functional rôle of the interstitial tissues is just as acceptable in the present state of knowledge as the sex-hormone hypothesis. Until experiments have been devised and carried out to dispose of one or the other alternative, it is reasonable to preserve an impartial attitude, bearing in mind William of Occam's aphorism *entia non multiplicanda praeter necessitatem*. Having employed this illustration, no further reason need be given for deferring the discussion of the rôle of the gonads in sex differentiation to the volume which deals specifically with that issue. There, for convenience, the endocrine significance of the ovarian follicles of the Mammal will also be set forth. But before proceeding to a consideration of the criteria for ascribing to an organ the function of internal secretion, as we have agreed to define it, there are, concerning the structure of the so-called ductless glands, some few particulars which will be assumed in the following chapters, and which may therefore be referred to most conveniently in this context.

§ II

The structures customarily referred to as ductless glands— pituitary, thyroid, parathyroids, suprarenals and "islets of Langerhans"—are exclusively confined to one phylum of animals, the Chordata, and within that phylum to one large group, the Craniata. It may be assumed that the reader is in a general way familiar with the histological characteristics and developmental origin of these organs. It will suffice therefore to draw attention to such particulars as emerge from recent work.

The practical justification for comparative physiology resides in the fact that some animals provide far more accessible material for the solution of a particular problem than others: given the solution of the problem in a particular species its general applicability is often less difficult to establish. As regards the suprarenal glands the anatomical relations of the constituent tissues are as favourable for experimentation in the Mammal as in any other Craniates; and a discussion of their histology will be reserved for the next chapter. A few remarks on the subject of the thyroid and pituitary glands are however necessary, partly because more is known of the physiology of these organs in the lower Vertebrates, and partly because in the case of the pituitary a good many new morphological facts have come to light during the past few years. A full account of recent work on the pituitary complex will be found in de Beer's monograph (1925). In the past it has been customary to distinguish two epithelial portions, the pars anterior (glandularis) and pars intermedia, derived from the ectoderm (hypophysis), from a portion (pars nervosa) which is non-glandular and derived from the infundibulum of the embryo. An important result of recent work is the recognition of a third glandular portion—the pars tuberalis. In view of the current confusion in nomenclature the following schema from de Beer's paper will be useful to our purpose:

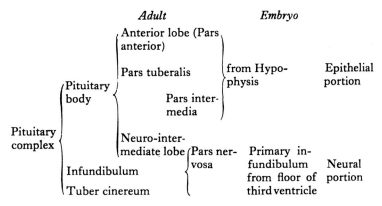

The pars tuberalis, which was first recognised in 1913 by Tilney and more extensively studied by Atwell and his co-workers, is present in all land Vertebrates with the possible exception of snakes and some lizards. In Mammals it lies dorsal to and in front of the pars anterior, set like a collar about the infundibular stalk and covered by the meninges, which fact renders it difficult to clear from the nervous tissue of the stalk itself. Easily recognisable even under low powers of the microscope by reason of its characteristic vesicular structure, the pars tuberalis can be very satisfactorily demonstrated by various staining methods. Though in contact with the pars intermedia in one part (hence formerly called the tongue-shaped process of the pars intermedia) the two portions have an entirely different ontogeny. In Birds the tuberalis is present in all the forms investigated (fowl, duck, pigeon, sparrow), and is specially interesting for reasons stated later. Among Reptiles the pars tuberalis is present in *Sphenodon*, Crocodilia and Chelonia, and is similar to that of Mammalia except that it does not completely invest the infundibular stalk and displays a less pronounced vesicular structure: it does not appear to be present in snakes and some lizards. The Anura present the most arresting feature from the experimental standpoint, since in these Amphibia the tuberalis consists of two lobes anterior to and completely detached from the rest of the gland. In Urodela the pars tuberalis is not as yet separated from the pars anterior. A comparison of the properties of extracts of the anuran gland with those of preparations of the posterior lobe of animals in which the tuberalis is incorporated in the latter may therefore be expected to throw light on the localisation of the active materials.

With regard to the supposed evacuation of the active substances from the pars intermedia into the pars nervosa two points are worthy of mention. It is first to be noticed that while the intermedia is almost devoid of blood vessels the pars nervosa, and especially that portion which borders on the pars intermedia, is well supplied with them in Mammalia, Amphibia and Teleostei. Secondly, though the pars intermedia and nervosa are prone to contain spherical cysts of hyaline material, this is not universal. In a number of cases, notably the cat among Mammalia, the occurrence of hyaline bodies is exceptional, and of fourteen cats the whole glands of which were cut into complete series of sections by de Beer, only one was found to possess these cysts. There are two points in which de Beer's interpretation of the structure of the pituitary in different vertebrate classes differs from that given by previous authors. First, with reference to the elasmobranch pituitary, the so-called inferior lobe has developmental similarities which recall the pars tuberalis of the Tetrapoda; but it is the existence of a pars intermedia and nervosa in the Selachians which concerns us here. From the fact that there is no well-marked lobe consisting of nervous tissue some authors have denied that there exists a differentiated pars nervosa in this group. There is, however, a mass, somewhat small it is true, of neuroglia fibres which leave the floor of the infundibular cavity and ramify in the intermedia. An increase in the quantity of this element would give rise to the condition in the haddock, for example, in which the nervosa is recognised

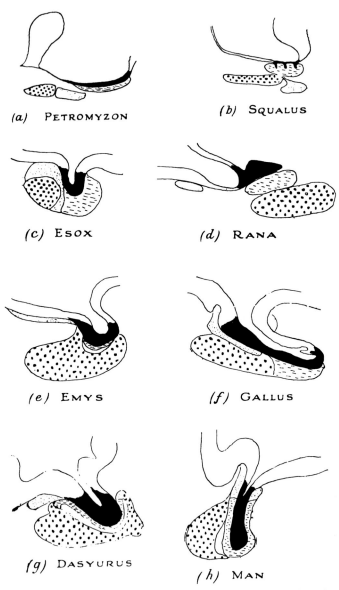

(a) PETROMYZON

(b) SQUALUS

(c) ESOX

(d) RANA

(e) EMYS

(f) GALLUS

(g) DASYURUS

(h) MAN

Fig. 1. Diagrammatic representation of the relations of the pituitary in the different groups of Craniates. Large dots, pars anterior; horizontal broken lines, pars intermedia; small dots, pars tuberalis (*übergangsteil*); black, pars nervosa; modified (c) and (e) from Stendell, (d) from Atwell, (f) and (h) from Tilney, (g) from Parker. The anterior end is to the left in each case.

to exist. A second point of some importance is that the pars intermedia, present in all other Craniates from *Petromyzon* to Man, does not seem to be separately differentiated in Birds, where the structure which has been labelled pars intermedia by some other writers appears for the following reasons to be pars tuberalis. It is a thin sheet of tissue encircling the infundibular stalk, but absent from the posterior end of the pars nervosa; it develops from the lateral lobes of the hypophysis; in development the tissue on the posterior side of the hypophyseal cavity is small in account, histologically identical with the developing pars anterior, and consequently, when the cavity becomes obliterated, there is no layer of tissue separating the pars anterior from the pars nervosa; finally, it is highly vascular and vesicular, the vesicles being of an altogether different character from the cysts sometimes found in the pars intermedia. It is well to emphasise one other point of some importance in comparing extracts from the several parts of the mammalian gland for experimental purposes. In some Mammals (pig, sheep, ox) there is a cone of tissue, first described by Wulzen, projecting from the anterior aspect of the pars intermedia. It varies greatly in size and may be larger than the rest of the pars intermedia, from which it differs considerably in histological appearance, resembling more closely the pars anterior with its dense granular eosinophil cells.

These points have been dealt with at some length, because it is quite impossible to regard the pituitary as a single organ from the physiological standpoint, and a clear appreciation of its anatomical complexity must be assumed in later chapters. For experimental purposes it is also valuable to bear in mind a few facts about the development of the gland. The origin of the glandular portion from a hollow ingrowth (Rathke's pouch) on the roof of the stornodalum in Selachians and Amniota is not a universal condition, a circumstance which is propitious to experimental procedure. The hypophysis is initially in Cyclostomes, Teleostei and Amphibia a solid ingrowth anterior to the mouth, so that at a certain stage of development (Fig. 2) it is easily accessible to injury or ablation. In fact the extirpation of the pituitary which has proved a stumbling block to the mammalian physiologist is a comparatively easy matter in Amphibia at all stages of development. In embryos of

Rana pipiens at the 3·5—4 mm. stage, the hypophyseal invagination is readily seen in front of the mouth between the protuberance

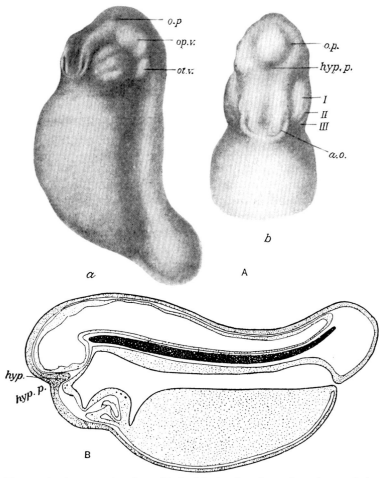

Fig. 2. A. A 4 mm. frog larva (*R. boylei*) showing the surface characteristics at a favourable stage for epithelial hypophysectomy; (*a*) ventrolateral, (*b*) ventro-cephalic view. B. A median sagittal section of a frog larva of approximately the same age and size as shown in figure A.

of the fore brain and the stomodaeum which is just forming. The epithelial ingrowth is removed with adjacent epithelium by means of a spear-point needle. Healing takes place within thirty minutes to three hours. P. E. Smith (1916) found that 60 per cent. of the operations were successfully accomplished without danger to the mouth rudiment. In adult Amphibia the pars nervosa is not connected by a solid stalk with the brain. Hence the gland (the pars tuberalis excepted) is easily removed, if a small hole is bored through the parasphenoid immediately below it by means of a dental drill with a rose burr. Through this perforation the gland can be sucked off without damage to adjacent parts by a fine glass tube connected with a filter pump. The location of the gland presents no difficulty. In the axolotl the operation takes less than two minutes if skilfully performed.

Turning now to the thyroid, the condition displayed by the *Ammocoetes* larva of *Petromyzon* leaves no alternative to the view that this gland which is found in all Craniates is structurally represented in the Protochordates by the endostyle. Like the pituitary its presence is restricted to the Craniata. In *Ammocoetes* an endostylar groove is present; but its communication with the pharynx is reduced to a duct. During metamorphosis the endostylar organ undergoes partial atrophy: three of its specialised types of epithelium—and the duct with them—disappear. The thyroid follicles of the adult, according to Marine, are formed from one persisting type of cell, the ciliated epithelial elements; and these cells sometimes retain their cilia after metamorphosis. Cowdry has described cilia in connexion with the glandular epithelium of the thyroid follicles in Selachians. Concerning the structure of the thyroid it is sufficient to say that in all Vertebrates it displays the same unmistakable features: it is made up of closely packed spherical or ovoid alveoli of cubical epithelium surrounding a mass of viscid globulin; and is very well supplied with blood capillaries. For experimental purposes it is important to bear in mind that thyroid tissue is one of the most labile tissues in the body: it is capable of rapid hypertrophy or involution.

The thyroid may remain single and median in position as in Elasmobranchs, or divide more or less completely into two lobes, which are widely separate from one another in Birds and Amphibia. For this reason the American workers have aimed at extirpating the median anlage in the frog embryo. In *Rana pipiens* the operation is recommended by Bennet Allen at the 6–7 mm. stage. An incision is made with a cataract needle transversely between the gland rudiment and the pericardium downwards towards the pharyngeal cavity. Care should be taken to remove the portions of the floor of the latter adjacent to the point from which the thyroid anlage is growing (Fig. 3). The wound should be visibly healed within half an hour. Concerning the comparative morphology of the parathyroids, which occur only in terrestrial Vertebrates, no comment is here necessary, since evidence bearing on the function of these organs is almost exclusively derived from experiments on Mammals.

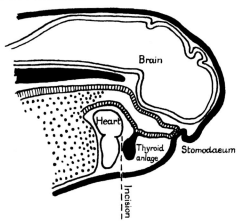

Fig. 3. T.S. head of frog embryo at stage when the thyroid rudiment is removed (Bennet Allen).

§ III

From this digression into morphological questions which need not be referred to again in detail we may turn to ask: what then are the criteria for ascribing to an organ the function of internal secretion? To answer this question, it is helpful to distinguish between two kinds of regulative processes: the regulation of specific effector responses, e.g. secretion of the pancreas or contraction of pigment cells, and the control of metabolic exchanges and developmental phenomena.

The first are in some ways more accessible to experimental treatment. If it is known that an organ contains a substance which evokes a specific local response in some effector unit (e.g. action of adrenaline on the pupil or of pituitary extract on amphibian melanophores) its endocrine function is sufficiently established by applying one of two methods. In the first place it may be shown that, when responses which can be evoked specifically in isolated effectors by its active material occur in the intact animal, they are associated with the liberation into the blood stream in amount significantly greater than that normally present of a substance having the same properties as its extract. This line of attack underlies certain researches on the mammalian suprarenals. One may formulate the alternative as follows. Given the fact that the organ yields an autacoid which evokes response in an isolated effector, it is also legitimate to conclude that such an organ is an endocrine organ when the consequences of removing it upon the given effector system are compensated by introducing the active substance into the circulation. Examples of this kind will be cited in dealing with the part played by ductless glands in regulating the chromatic function of Reptiles and Amphibia. When it is not possible to isolate the seat of a response which depends upon the presence of an organ or is evoked by injection of its extracts, it is not so easy to interpret the facts with certainty, even though the effects of administering its extracts can be brought into harmonious relationship with the consequences of experimental extirpation. Adrenaline has a local action on the melanophores (black pigment cells) of Reptiles, Amphibia and Fishes. If, as Redfield describes, excitement pallor can no longer take place after removal of the adrenals, or locally, after blocking of the blood supply of a limb, there is, in the light of the established effect of adrenaline in promoting contraction of the melanophores, sound justification for the belief that adrenal secretion is a pre-eminent factor in the production of excitement pallor in Reptiles. But it is not so clear from effects produced by injection of insulin or thyroxine after extirpation of the pancreas or thyroid that these autacoids are ordinarily set free in the blood stream—the possibility that insulin normally produces its effect on the blood sugar *in situ* is excluded by evidence from other sources.

The criteria which have been laid down have only been satisfied in a small number of cases. But when it is remembered that the mechanism of internal secretion was not clearly established before the work of Bayliss and Starling in the opening years of the present century, it is no matter of surprise that our knowledge is still fragmentary. For this reason one cannot avoid dealing with much that is still hypothetical. And in doing so selection must be made from an immense literature dealing with observations that are by no means unanimous. There are few so-called ductless glands whose removal has not been correlated by some observer with retardation or stimulation of sexual development; and it is less surprising to notice the number of tissues whose extracts have been described as possessing depressor-diuretic properties. In confining the account given in the pages which follow to what the writer regards as some of the more suggestive lines of enquiry relevant to the influence of internal secretion in the regulative processes of the organism, it is taken for granted that the reader has access to such authoritative and exhaustive treatises as Professor Sharpey Schafer's monograph on the Endocrine Organs.

REFERENCES

DE BEER. *The Comparative Anatomy etc. of the Pituitary.* Oliver and Boyd, Edinburgh. 1925.

BENNET ALLEN. *Journ. Exp. Zool.* 26, 1918.

MARINE. *Journ. Exp. Med.* 17, 1913.

SMITH, P. E. *Anat. Mem.* 11, 1921.

Chapter II

ADRENALINE AND NEUROMUSCULAR CO-ORDINATION

§ I

In the chapters which follow we shall have to deal exclusively with the activity of organs which are not yet known to have any precise parallel among the invertebrate phyla, organs whose products, moreover, are not as yet known to elicit any definite physiological responses in the tissues of animals other than Craniates. The case of the so-called chromaphil or chromaffine cells and their active product, adrenaline, is an exception. There is, as will be seen, a certain amount of evidence that adrenaline-secreting tissue is widely distributed throughout the animal kingdom; and there is sufficient justification for the statement that adrenaline and allied chemical substances excite the muscles of Invertebrates as well as Vertebrates. It is therefore appropriate that we should commence our survey of the comparative physiology of internal secretion with a consideration of adrenaline and its relation to neuromuscular co-ordination.

Adrenalin(e), also called epinephrin (Abel), suprarenin (von Furth) and adrenine, is the name given by Takamine to the natural base first isolated by him in 1901 from the substance of the medullary portion of the mammalian suprarenal gland, in which it is present according to determinations of Elliott (1912) and of Folin, Cannon and Denis (1913) in amount varying from 0·1 to 0·3 per cent. Adrenaline possesses all the characteristic properties exhibited under experimental conditions by extracts of the suprarenal medulla of the Mammal and analogous tissues of other Vertebrates. The balance of evidence is in favour of the view that in Mammals it is actually liberated into the blood stream, and that its appearance in the circulation of the animal may be correlated with definite responses, i.e. that the suprarenal medulla of the Mammal is an organ of internal secretion in the sense defined in the preceding chapter. This view based largely on the work of

Elliott and of Cannon and his collaborators has not however escaped criticism. But before entering upon a discussion of the available data, it will be necessary to give a brief account of the physiological properties of adrenaline, its chemical structure and the issues arising therefrom. Finally, attention will be directed to the phyletic distribution of adrenaline, and the possibilities of further research in the comparative physiology of internal secretion suggested by what is already known. The relation of adrenaline-secretion to colour response in the lower Vertebrates will be dealt with in a later chapter.

As the suprarenal gland of the Mammal is a composite structure whose elements are somewhat differently distributed in other Vertebrates, it is more satisfactory for purposes of comparative physiology to refer to the cells in which adrenaline is formed as "chromaphil tissues."

§ II

Study of the physiological properties of extracts of the chroma-phil tissues began when Oliver and Schafer (1894) made their historic researches on the effect of a variety of tissue extracts (including that of the pituitary) on the blood pressure of the Mammal; and demonstrated the remarkably powerful pressor effect obtained from preparations of the suprarenal medulla on intravenous injection. This observation, which may be regarded as the starting point of modern research in the field of internal secretion on experimental lines, was made shortly after and independently by Cybulski (1895) and by Szymonovicz (1895). The rise in blood pressure begins within a few seconds after injection; and is in the main due to arterial constriction: isolated portions of artery have been shown to contract in the presence of adrenaline by numerous workers. The response is common to all land Vertebrates: it has not as yet been demonstrated in fishes. Among Amphibia it has been described in the frog by Kuno (1918) and by Hogben and Schlapp (1924) from observations on the pulmocutaneous artery; while constriction of the vessels by perfusion is now included in most practical courses in elementary physiology. Among Reptiles the pressor response has been shown to occur in various species of Chelonia by Edwards (1914) and Hartman, Kilborn and Lang

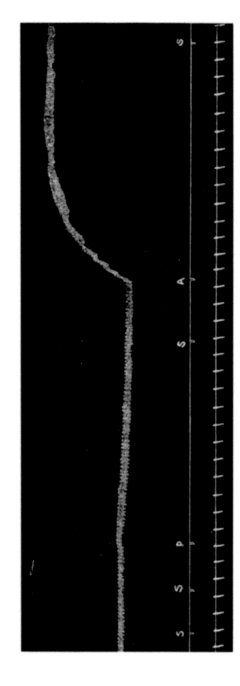

Fig. 4. Tortoise, subclavian blood pressure. In order from left to right: 1 c.c. saline; 1 c.c. saline; 10 mg. sample B pituitary extract; 1 c.c. saline; 0·07 gr. suprarenal extract (Armour).

(1918). In both cases the threshold is much higher than in warm-blooded Vertebrates. Hogben and Schlapp (1924) record observations which suggest that adrenaline has a less pronounced effect than the allied synthetic base epinine (cf. §§ iii and v) on the blood pressure of the tortoise, as is also the case with certain invertebrate tissues. The reverse is true of the mammalian response. The blood pressure of the Bird, like that of the Mammal (Hartman, Kilborn and Lang), is much more sensitive to adrenaline than that of the Reptile. But in Birds, as in Reptiles, purely constrictor effects are obtained with intravenous injection of the autacoid. In Mammals, on the other hand, very small doses of adrenaline often elicit a pure depressor effect. Thus in the opossum (Hartman, Kilborn and Lang) 0·2 c.c. of a 1 : 100,000 solution of adrenaline evoked a pure fall, while 0·5 c.c. of a 1 : 10,000 solution called forth the characteristic pressor response. The same authors describe depressor effects with quantities of adrenaline below the threshold for the pressor response in various genera of Mammals (raccoon, goat, rabbit, monkey). The mechanism of this fall is not fully understood; but it seems likely from recent work of Krogh and of Dale and Richards (1919) that it may be due to local dilation of capillaries by adrenaline. It is interesting to note that, as with the pressor response to pituitary extract, tolerance is established to successive doses of the order which evoke the depressor effect. Dale (1906) showed that after injection of ergotoxine, adrenaline produces in the Mammal a steep fall in doses which normally evoke a pressor effect. Ergotoxine abolishes motor effects of adrenaline upon plain muscle, and it is possible that in the reversal which Dale discovered, the action of the drug is to paralyse the excitatory mechanism of the arterioles, thus unmasking the inhibitory action of the autacoid on certain of the capillaries. However, the reversal of the characteristic activity of adrenaline can also be witnessed in isolated effector units, as is evident from the experiments of Spaeth (1916) on the melanophores of the Atlantic minnow, *Fundulus*. Spaeth and Barbour found that adrenaline and various allied synthetic amines (phenylethylamine, oxyphenylethylamine, indolethylamine) bring about contraction of the pigment cells of *Fundulus*, the threshold dilution for the natural autacoid being of the order 1 : 50,000,000. If scales

are immersed for a time in a solution of ergotoxine before being
placed in a solution of adrenaline they respond in the opposite

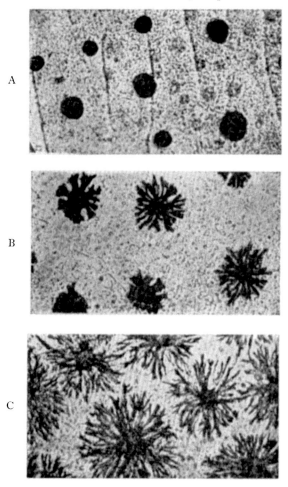

Fig. 5. Melanophores of *Fundulus heteroclitus* (Spaeth and Barbour). A, effect
of adrenaline; B, of ergotoxine alone; C, of adrenaline after ergotoxine.
Reproduced from Schafer's *Endocrine Organs* by permission.

manner, i.e. by maximal expansion. It seems more likely in a case
like this that something analogous to the action of strychnine on
the nerve cell is involved.

The pressor response to successive equivalent doses in the same animal under properly controlled conditions is identical; and as the rise rapidly subsides the pressor phenomenon is admirably suited to quantitative estimations where the amounts to be compared are not too small. The pressor assay of adrenaline is due to Elliott (1912), who matches the effect of successive doses on the carotid blood pressure of the spinal cat; and obtains in this way a greater accuracy than is usually obtainable by colorimetric methods.

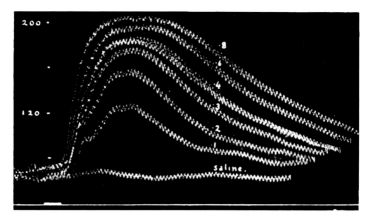

Fig. 6. Estimation of the relative amount of adrenaline in different extracts by comparing the effect of the same quantity of each upon the blood pressure of a pithed cat (T. R. Elliott). ·1, ·2, etc., represent the effect of successive doses containing o·1, o·2, etc., milligrams adrenaline chloride. At ·4 a double tracing is noticeable, this amount having been repeated after ·8.

In their original contribution on the action of suprarenal extracts Oliver and Schafer also recorded the observation that the active principle not only induces arterial constriction, but produces augmentation and acceleration of the mammalian heart after section of both vagi. This effect is seen in the isolated heart in which Evans and Ogawa (1914) have studied its mechanism. Adrenaline greatly increases the total gaseous exchanges of the heart, the oxygen consumption increasing roughly in proportion to the pulse rate, without any increase or diminution of the power of the tissues to utilise carbohydrate or change in the mean respiratory quotient. In Mammals both auricles and ventricles are affected. In the Bird (pigeon) and Reptile (tortoise), according to

Elliott (1905), only the auricles are excited by adrenaline. In the case of the Bird, Paton (1912) confirms Elliott's findings that adrenaline does not excite the ventricles; but records his inability to obtain any augmentor response from the intact auricle. There is no augmentor nervous mechanism in connexion with the avian ventricle (Paton). Several observers have described the augmentation and acceleration of the isolated heart of the frog with adrenaline; but information under this heading has not always been

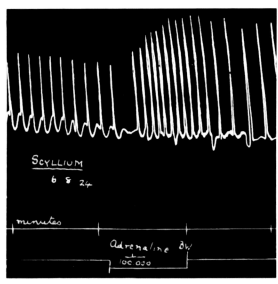

Fig. 7. Response of heart of dogfish to adrenaline, 1 : 100,000 at signal.

safeguarded with proper attention to the hydrogen ion concentration of the medium. Recently Macdonald (1925) has studied the action of adrenaline on the perfused heart of the Elasmobranch. The phenomena described by him are puzzling. He emphasises as the characteristic effects on the cardiac rhythm of the dogfish— (a) an initial and often very striking inhibition in the normal rate, (b) subsequent increase in the amplitude which may be but is not invariably associated with acceleration, (c) an increased tolerance to hydrogen ions. The last is suggestive as a basis for further investigation of the physical nature of the mechanism through which adrenaline operates.

The pharmacodynamic properties of adrenaline first recognised in connexion with the circulatory system of the Mammal were studied in relation to other contractile mechanisms by Lewandowsky (1898–1900) and Boruttau (1898), who showed that intravenous injection of suprarenal extract produces dilatation of the pupil, inhibition of the bladder, and intestinal movements and excitation of the arrectores pilorum. Langley (1901) extended these observations, in particular drawing attention to the action of adrenaline on the salivary glands and genital ducts. Langley first emphasised the fact that "the effects produced by suprarenal

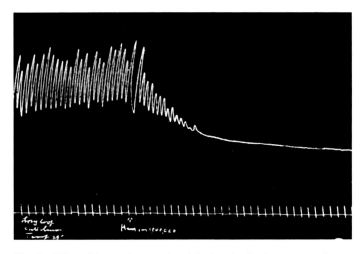

Fig. 8. Effect of immersing a strip of the longitudinal coat of the ileum of a cat in 1:1,000,000 solution of adrenaline (A. W. Young).

extract are almost all such as are produced by stimulation of one or other sympathetic nerve," and "in no case corresponds to that which is produced by stimulation of a cranial autonomic or a sacral autonomic nerve." To the former conclusion, however, Langley himself furnished an important exception, namely, the failure of adrenaline to excite the sympathetically innervated sweat glands of the Mammal. Later, Langley (1903) showed that injection of adrenaline produces erection of the feathers in the bird, a phenomenon also evoked by sympathetic stimulation.

These discoveries led to a closer study of the action of adrenaline on plain muscle, so making available for physiological investigation specific biological indicators of much greater sensitivity than the pressor response. Of these, inhibition of the tone and rhythm of the mammalian intestine is of special importance. That complete inhibition can be obtained in a strip of rabbit's intestine with dilutions of 1 : 1,000,000 is a commonplace of laboratory classwork. Magnus (1905), who specially developed the technique, records inhibition of the longitudinal coat of the cat's ileum with dilutions of 1 : 20,000,000. Hoskins obtained inhibition of the rabbit's

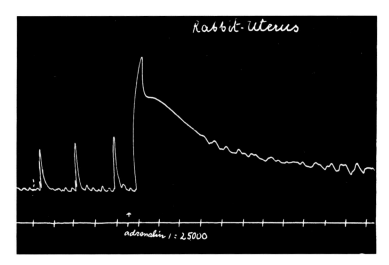

Fig. 9. Effect of solution of adrenaline on cornu uteri of rabbit, the contraction of which is increased (Itagaki).

intestine at a dilution of 1 : 400,000,000, and Stewart and Rogoff (1912) with 1 : 800,000,000. The reaction of the uterus is of special interest in devising specific tests of high sensitivity, because the character of the effect depends on the species of animal and its age. The uterus of the rabbit is said to contract in dilutions of 1 : 20,000,000 adrenaline. This is the more common type of response. But the uterus of some Mammals (rat, guinea-pig, mouse), like the intestine, is inhibited by adrenaline (cf. Itagaki, 1917). As shown by Dale (1906) and Cushny (1906), the non-pregnant uterus of the cat is inhibited, while the virgin uterus contracts in the presence of adrenaline. The pupillary reaction is also of great sensitivity. The enucleated eye of the frog when placed in a solution of adrenaline in Ringer's fluid dilates at a dilution of 1 : 20,000,000. As pointed out below, degenerative section of the sympathetic nervous supply increases the reactivity of muscle to the autacoid. An important advance in the technique of detecting adrenaline in very small quantities was made by Cannon (1917), who introduced the denervated mammalian heart as an indicator of adrenaline.

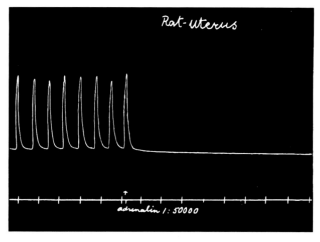

Fig. 10. Effect of solution of adrenaline upon an isolated portion of uterus of rat (Itagaki). The contractions are inhibited.

The discovery by Dixon (1904) that apocodeine abolishes response to adrenaline as well as to the stimulation of sympathetic nerves has, in the light of Langley's generalisation, given rise to a good deal of discussion as to the seat of action of the autacoid. Since it has been repeatedly shown (Lewandowsky, Langley and Elliott) that degenerative section of the postganglionic neurone does not abolish the response to adrenaline, even increasing the sensitivity of the muscle (Elliott) to the latter, it is clear that adrenaline acts on some mechanism peripheral to the visible nerve endings. To the general parallelism between the effects of stimulating sympathetic nerves and the action of adrenaline on the structures innervated there are two noteworthy exceptions, the failure of adrenaline to excite the sweat glands of the Mammal, and the constrictor action of adrenaline (Carlson and Luckhart, 1919) on the pulmonary circulation of the frog and turtle, an effect not produced by stimulating any sympathetic nerves. Whatever the significance of this parallelism may be it is doubtful if further discussion of the problem is likely to prove profitable till we have at our disposal more quantitative data based on methods for the study of excitatory processes such as have been employed by Lapicque and Keith Lucas.

One characteristic response produced by introduction of adrenaline into the circulation has yet to be mentioned, namely, the glycosuria, as first shown by Blum (1901), which occurs in the Mammal. This effect is a persistent one—it may last for several days—and cannot therefore be attributed to vasomotor disturbances. Little is known of its mechanism, except that it is directly related to the glycogenic function of the liver and is apparently independent of the activity of the pancreas.

The responses which have been dealt with in the foregoing remarks are evoked by quite minute quantities of adrenaline. But comparatively small doses suffice to bring about death in Mammals. Thus intravenous injection of a quantity of adrenaline equivalent to about a tenth of a milligram per kilo body-weight is lethal in some cases; and even with subcutaneous injection 10 mg. is the lethal dose for a guinea-pig. This is not without interest from the standpoint of comparative physiology, partly because of the occurrence of adrenaline and allied compounds in the poison glands of certain animals and partly because of the powerful physiological action of adrenaline on some of the lower forms on which they prey. Adrenaline is present in the secretion of the two large oval "parotid" glands and in certain of the cutaneous glands of the toad *Bufo agua*. Its presence was demonstrated by Abel and Macht (1912) along with an alkaloid *bufagin* which resembles digitalin in some of its pharmacodynamic properties, increasing the tonicity of the heart and producing marked diuresis. An allied compound exists in the salivary secretion of Cephalopods, and enables them to paralyse their crustacean prey. The action of the salivary juice of Cephalopods (*Eledone, Sepia, Loligo* and other genera) has been studied, especially by Livon and Briot (1913). Injection of 0·5 c.c. produces general paralysis in the crab within two minutes; and the animal eventually dies. The paralysis is accompanied by failure of the appendicular muscles to respond to electrical stimulation and cessation of the heart beat. Henze (1913) studied the chemical properties of the salivary juice and came to the conclusion that the active substance was parahydroxyphenyl-ethylamine (tyramine), one of a series of compounds allied to adrenaline, prepared synthetically by Barger and Dale. The same substance has been stated by Boruttau and Cappenberg to occur in Shepherd's Purse (*Capsella*). Botazzi (1919) has recently questioned Henze's identification of the salivary poison of Cephalopods with tyramine on the grounds that the substance has a different mode of action on plain muscle.

REFERENCES

ABEL and MACHT. *Journ. Pharm. Exper. Ther.* 3, 1912.

BOTAZZI. *Pubbl. Staz. Zool. Napoli*, 1919.

CUSHNY. *Journ. Physiol.* 35, 1906.

DALE and RICHARDS. *Ibid.*

EDWARDS. *Am. Journ. Physiol.* 33, 1914.

ELLIOTT. *Journ. Physiol.* 32, 1905; 44, 1912.

HARTMAN, KILBORN and LANG. *Endocrinology*, 2, 1918.

HENZE. *Zeit. Physiol. Chem.* 87, 1913.

HOGBEN and SCHLAPP. *Quart. Journ. Physiol.* 14, 1924.

ITAGAKI. *Quart. Journ. Exp. Physiol.* 9, 1917.

LANGLEY. *Journ. Physiol.* 27, 1901.

LIVON and BRIOT. *Journ. de Physiol. et de Pathol. Gen.* 8, 1913.

MACDONALD. *Quart. Journ. Physiol.* 15, 1925.

OLIVER and SCHAFER. *Journ. Physiol.* 27–8, 1894–5.

PATON. *Journ. Physiol.* 44, 1912.

SPAETH and BARBOUR. *Journ. Pharm. Exp. Ther.*

§ III

Many years before the physiological action of suprarenal extracts had been established by the classical researches of Oliver and Schafer, Langley and others, the clinical observations of Addison (1849) and of Brown-Sequard (1856) had led several

workers to study their chemical properties. Vulpian (1856), whose
work was confirmed by Virchow a year later, first showed that the
medullary portion of the mammalian gland yields a specific substance which is coloured green by ferric chloride and rose-red by
iodine in solution. During the latter half of the nineteenth century
observations of a similar nature were made by Arnold (1866),
Holm (1867), Krukenberg (1885) and Brunner (1892). A new
phase in the biochemistry of adrenaline began when Moore
(1895–7), working in Schafer's laboratory, correlated the pressor
activity of suprarenal extracts with the intensity of the colour
reactions of Vulpian, thus encouraging the belief that the active
substance would be found to be identical with Vulpian's chromogen.
In the next few years numerous attempts to isolate the substance
responsible for the colour tests were made, notably by von Furth
(1898–1901) and Abel (1897–1901). The former obtained an exceedingly active preparation by precipitation with ferric chloride
and subsequent purification by solution in methyl alcohol; and
Abel succeeded in preparing a benzoyl derivative. In 1901
Takamine announced the isolation of "adrenaline" by a method
which consisted in concentrating the acidified aqueous extract by
evaporation, precipitating the remainder with alcohol, evaporating
the filtrate *in vacuo*, and redissolving in ammonia, ammonium
chloride or sodium hydrate. The crystals which separated from
the alkaline solution were purified by recrystallisation and found
to be homogeneous. Aldrich independently in the same year
obtained a chemically pure product, to which he correctly assigned
the empirical formula $C_9H_{13}O_3N$. Aldrich proceeded by a method
not very different from that of Takamine; but before alcoholic
precipitation of the concentrated extract, he added neutral lead
acetate, centrifuging and afterward removing excess of lead with
sulphuretted hydrogen. From its solution in benzaldehyde Barger
and Ewins (1906) found the molecular weight of adrenaline to be
170 at 90° C.

The chemical constitution of the compound was soon established.
Takamine himself showed that it was a catechol derivative; and
von Furth demonstrated the presence of a methylamino group
and a hydroxyl in the side chain. Pauly showed that it possessed
an assymetric carbon atom and two hydroxyl groups attached to

the nucleus, thus reducing the possible constitutional formulae to two:

(1)

OH

OH

CHOH

CH$_2$NHCH$_3$

(2)

OH

OH

CH . NHCH$_3$

CH$_2$OH

Friedmann (1904) showed that natural adrenaline is a secondary alcohol, as indicated by the first formula, synthesised by Stolz (1903) and Dakin (1905). The identity of adrenaline with Pauly's formula was finally established by Flächer (1908).

The chemical and physical properties of adrenaline which are fully set forth in Barger's monograph on *The Simpler Natural Bases* may be briefly summarised as follows. It is a fairly strong base, slightly soluble in water in the cold—0·0268 per cent. at 20° C.— rather more soluble in boiling water, insoluble in most organic solvents, but readily dissolving in the theoretical quantity of a mineral acid, in glacial acetic, warm ethyl oxalate and in benzaldehyde. Being a phenol, it is also soluble in caustic alkalis but not in ammonia and sodium carbonate solution. Solutions of adrenaline assume a reddish hue immediately with mild oxidising agents; and oxidation with the production of the characteristic coloration takes place on exposure to air, though more rapidly in alkaline than in acid solution, especially if minute traces of iron are present. Natural adrenaline is laevorotatory: *d*-adrenaline, as was first shown by Cushny, is much less physiologically active with respect to those responses which have been studied from this point of view. Crystalline hydrochlorides, bitartrate, oxalate and urate have been prepared, as also an amorphous borate. The colour reactions are of importance in physiological as well as biochemical investigation. These are notably: (*a*) the grass green tint in presence of ferric chloride turning to violet and red on addition of dilute alkali, given at dilutions of 1 : 30,000; (*b*) the pink hue given on exposure to air or in presence of mild oxidising agents such as oxidases, iodine (1 : 1,500,000), mercuric chloride, potassium persulphate (1 : 5,000,000), and potassium ferricyanide; (*c*) a reaction with sodium tungstate (Folin, Cannon and Denis) also given by uric acid is capable of detecting adrenaline at a dilution of 1 : 3,000,000. These methods have been used as a basis of quantitative work by colorimetry.

Numerous bases structurally related to adrenaline have been prepared synthetically, one of which has been mentioned already. The three most closely allied to the natural autacoid in their biological properties are:

3 : 4 dihydroxy-phenyl-ethanolamine	$(OH)_2.C_6H_3.CHOH.CH_2.NH_2$ "arterenol"
3 : 4 dihydroxy-phenyl-ethyl-methylamine	$(OH)_2.C_6H_3.CH_2.CH_2.NH.CH_3$ "epinine"
o-ethyl amino-3 : 4 dihydroxy-acetophenone	$(OH)_2.C_6H_3.CO.CH_2.NH.C_2H_5$ "homorenon."

An extensive study of the physiological properties of amines more or less closely allied to adrenaline has been made by Barger and Dale (1910) who employ the term sympathomimetic to indicate the general similarity of their physiological action to responses evoked by sympathetic stimulation. Catechol has no sympathomimetic action: on the other hand sympathomimetic action is evoked by a large series of primary and secondary amines including some of the primary aliphatic amines. On the whole the intensity of sympathomimetic activity increases as the structure approaches more closely to that of adrenaline. The optimum conditions of chemical structure for physiological efficiency are: (a) a carbon skeleton consisting of a benzene nucleus with a side chain of two carbon atoms to the terminal member of which the amino group is attached, (b) phenolic hydroxyls in the 3 : 4 position, and if this condition is fulfilled, an alcoholic OH further intensifies activity. Motor and inhibitory action are not precisely parallel.

As to the origin of adrenaline in intermediate metabolism little is known: it does not seem unlikely that tyrosine is a parent substance. Elliott found that at birth there is practically no adrenaline in the suprarenal gland of Man. Fenger (1912) on the other hand was able to detect its presence in the suprarenal of the foetus of the pig, ox and sheep within a few weeks of conception; and Hogben and Crew (1923) obtained positive reaction of the mammalian gut to extracts of chromaphil tissue from chick embryos at an early stage of incubation.

REFERENCES

ALDRICH. *Am. Journ. Physiol.* 5, 1901.
BARGER and DALE. *Journ. Physiol.* 44, 1912.
CUSHNY. *Journ. Physiol.* 37–8, 1908–9.
ELLIOTT. *Journ. Physiol.* 44, 1912.
FENGER. *Journ. Biol. Chem.* 1912.
FLACHER. *Zeit. Physiol. Chem.* 58, 1908.
FOLIN, CANNON and DENIS. *Journ. Biol. Chem.* 12, 1912.
HOGBEN and CREW. *Brit. Journ. Exp. Biol.* 1, 1925.
MOORE. *Journ. Physiol.* 17, 1895; 21, 1897.
TAKAMINE. *Ibid.* 27, 1901.
VULPIAN. *Comptes Rendus,* 43, 1856.

§ IV

In the previous chapter we have sought to indicate the nature of the evidence which is necessary for a satisfactory proof of internal secretion as therein defined. In considering the physiological significance of adrenaline in the normal activities of the organism the data available with the exception of those discussed in the succeeding chapter are derived exclusively from experiments on the Mammal, and in Mammals attempts to remove the suprarenals have shed no light on the problem. Epinephrectomy is fatal in Mammals. The fatal consequences of the operation would however seem—apart from the question of shock, which has been an important factor in many recorded cases—to be due to the removal of the cortex rather than to removal of chromaphil tissue. In any case the anatomical relations of the latter in the Mammal have proved so far an insuperable barrier to the attempt to obtain any very definite information on the function of the medulla. Since, for the reasons stated, it is not possible to bring the results of extirpation into harmonious relationship with the properties of the active constituent of the chromaphil tissue, it is necessary to fall back on the attempt to correlate normal modes of response analogous to those evoked by injecting adrenaline into the circulation with the presence in the blood of a substance which has the same properties as the autacoid. To separate what may presumably be regarded as fairly well-established from more questionable issues it will assist us to consider, first, the more general question as to whether adrenaline is in fact liberated into the circulation at all;

secondly, whether under any circumstances the liberation of adrenaline can be stimulated; and thirdly, whether there is any evidence that any normal activities of the organism can be correlated with the activity of the chromaphil tissues.

In attempting to detect the presence of adrenaline in the blood the most useful indicator is the reaction of isolated gut. This test combines a high degree of sensitivity with a high degree of specificity. Very soon after the publication of Oliver and Schafer's discovery several investigators claimed to detect adrenaline in the blood leaving the suprarenals. It is now certain that these observations were worthless. The careful quantitative work which has been carried out since the elaboration of the intestinal and pupillary (denervated) reactions and the variety of tests showing considerable gradation of delicacy leaves no doubt that there is adrenaline in the blood of the suprarenal veins; and that it is never present in quantities that can be measured by such gross procedure as that employed by the pioneer workers in this field. The blood of the suprarenal veins gives positive intestinal and pupillary reactions which are not given by the blood from the general circulation. The admirable quantitative work of Stewart and Rogoff (1923) has made it possible to get a clear idea of the magnitude of the normal output of adrenaline in dogs and cats. For 103 cats the average output per minute per kilo body weight was found to be 0·000226 ± 0·000007 mg. of adrenaline. For 32 dogs the average was statistically identical, viz. 0·00027 ± 0·0000096. These figures are based on the intestinal test. Control estimations of the pupillary and intestinal reactions carried out with blood from the adrenal veins in seven cases (cats) gave values of the same decimal order of magnitude in each case:

No. of in-individual used in experiment	Output in mg. per minute per kilo body wt.	
	Pupillary	Intestinal
85	0·0003	0·00024
86	0·0007	0·00014
814	0·0002	0·00029
815	0·0003	0·00053
816	0·0003	0·00037
817	0·00015	0·00018
821	0·0004	0·00076

The normal output of adrenaline in cats and dogs according to Hoskins is of about the same order as the minimal quantity requisite to produce perceptible vasodilator effects: it is far below the threshold for pressor action. The likelihood that adrenaline-secretion plays any important part in vasomotor regulation may now be dismissed.

Several attempts have been made to show indirectly that the discharge of adrenaline can be accelerated by nervous stimulation. Elliott (1912) found, for instance, that on stimulating the splanchnics a triphasic pressor (carotid) response is ordinarily obtained: there is an initial rapid rise followed by a distinct depression which

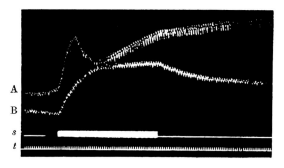

Fig. 11. Effects on blood pressure of stimulating the splanchnics in cat before (A) and after (B) removal of suprarenal capsules (T. R. Elliott). A, immediate rise, with dip, followed by a prolonged rise; B, immediate and prolonged rise, without dip; s, signal; t, time, in seconds.

gives place to a slow ascent to or above the initial rise. On ligaturing the vessels supplying the suprarenal glands the response is monophasic. This circumstance is readily explicable on the assumption that the first phase is due to direct nervous influence, the second as the initial outpouring of adrenaline in amount necessary to produce vasodilation, and the third a combined effect of nervous action and adrenaline discharge in sufficient quantity to exert a pressor action. Though Elliott's interpretation of the phenomena has been called in question by Gley and Quinquaud (1921), the confirmatory experiments of Vincent and Pearlman (1917) and of Vincent and Wright (1924) afford some justification for the belief that direct stimulation of the splanchnic nerves may result in the liberation of adrenaline into the circulation in quantity sufficient to evoke manifest physiological effects.

While there is not complete unanimity as to the possibility of stimulating the secretion of adrenaline by nervous agencies there is a good deal to be said for the affirmative view. As to whether the activation of the adrenals can be correlated with normal response of the organism, quite divergent views have been advocated by Cannon and his co-workers on the one hand and Stewart and Rogoff on the other. The reader who wishes for more detailed information regarding this controversy may refer to the recent reviews written respectively by G. N. Stewart and R. G. Hoskins.

The possibility that those motor reactions characteristic of sympathetic nervous stimulation occurring normally in fright might be augmented and prolonged by adrenal secretion was first investigated by Cannon and de la Paz (1911). Using a flexible catheter which was introduced through the femoral vein into the postcaval at the level of the lumboadrenal vessels, they withdrew blood from cats before and after being subjected to excitement, and obtained positive intestinal reactions from blood of excited animals. They failed to obtain the reaction after excision of the adrenals. Cannon and Hoskins (1911) employing the same type of procedure concluded that adrenaline is discharged in increased quantity during asphyxia or stimulation of sensory nerves (e.g. central end of the sciatic).

To do justice to both points of view it will be best to quote from Stewart and Rogoff's detailed criticism of these experiments the following passage which sums up their main line of argument:

As a method of estimating changes in the rate of output of epinephrin from the adrenals the catheter method is defective in principle. For at best all that could be measured by it would be changes in epinephrin concentration in the blood of the inferior cava above the level of the adrenals. Changes in the rate of flow of blood are not taken account of. The quantity of epinephrin flowing along the inferior cava to the heart per unit of time cannot be estimated, nor the changes, if any, produced in this quantity by the conditions studied. We have pointed out that in the cases in which Cannon claims to have obtained evidence of the increased output of epinephrin, all he can possibly have observed is an increased concentration in the caval blood, and that a slowing of the caval flow would cause an increase in the concentration, if no change whatever had occurred in the rate of output. We have suggested certain factors associated with all the conditions studied by him which might cause such changes in the caval flow, as could tend to increase the epinephrin concentration even in the absence of an increased rate of output.

A new line of attack on this problem was introduced by Cannon (1917) when the denervated heart was proposed as an indicator. And Cannon and Rapport (1921) showed that asphyxia or sciatic stimulation increases the rate of the denervated heart even when the possible changes in concentration through redistribution are reduced to a minimum by ligating the carotids, brachials and aorta below the level of the adrenal arteries. This would at first sight appear to be conclusive. But Stewart and Rogoff subsequently obtained the same result after ligature of the adrenal vessels. The same authors also claim that the "fright" manifestations described by Elliott in dogs under the action of morphia occur when the adrenals are excluded from the circulation. In a later communication Cannon and Uridil (1921) report that splanchnic stimulation sometimes produces increased cardiac rhythm after excision of the stellate ganglion and section of the vagi, but not after the hepatic nerves are cut. When this precaution is taken the exclusion of the adrenals always abolishes the response of the denervated heart to asphyxia and sciatic stimulation.

Until the possibility of response in the denervated heart to agencies other than adrenal secretion are more fully understood, it is impossible to accept without reservation the evidence derived from this source in confirmation of Cannon's earlier experiments. Perhaps the present status of the problem can be best summed up by saying that there are sufficient grounds to justify the conclusion that adrenaline is set free into the blood stream; that it is neither unlikely nor proven that increased secretion of adrenaline plays a minor part in reinforcing the motor phenomena associated with fright and asphyxia; and there is no reason to believe that in the Mammal the chromaphil tissue serves a prominent or indispensable function in regulating the normal activities of the organism.

REFERENCES

For complete bibliography consult:

CANNON. *Bodily changes in pain, hunger, fear and rage.* Appleton, New York, 1915.

HOSKINS. *Physiological Reviews*, 2, 1922.

STEWART. *Ibid.* 4, 1924.

§ V

It was first discovered by Henle (1865) that the cells of the suprarenal medulla in the Mammal give a characteristic reaction with chromic salts. This chrome-staining property is also found by groups of cells elsewhere in the body, though mainly confined to the suprarenal medulla in the Mammal; and it has proved serviceable in tracing the structure associated with the production of adrenaline in other groups of animals. The mammalian gland consists of components which have as far as we know no direct connexion physiologically and are morphologically quite distinct entities. The chromaphil cells arise in development in association with groups of sympathetic ganglion cells. They occur in close association with the latter in the adult of the lower vertebrates; and Elliott (1913) has brought forward evidence to support the conclusion that the adrenaline-secreting cells of the Mammal are innervated by medullated preganglionic fibres like sympathetic ganglion cells and not by postganglionic neurones, as are other glands. Nothing is known of the function of the suprarenal cortex of the Mammal which would justify discussion in these pages.

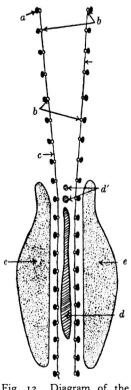

Fig. 12. Diagram of the suprarenal apparatus of an elasmobranch fish (Swale Vincent, *Internal Secretions*). *a*, paired chromaphil bodies (medullary substance) attached to *b*, sympathetic ganglia; *c*, sympathetic chain; *d*, interrenal body (cortical substance); *d'*, accessory interrenals; *e, e*, kidneys.

For descriptive purposes it is best to refer always to the tissues homologous with the mammalian suprarenal medulla as chromaphil tissues. Composite suprarenals are found in Birds and Reptiles, but the cortical and chromaphil elements form an interlacing system without a clear regional demarcation. Islets of cells of both types occur in juxtaposition scattered over the surface of the kidney in Amphibia, tending to form a more or less compact body in the Anura. But groups of chromaphil cells are also found within the sympathetic ganglia of these animals. The arrangement is much more diffuse in Fishes, where the disposition of the chromaphil tissues has been studied by histological and experimental methods by Giacomini and by Swale Vincent. Balfour (1878) suggested that the suprarenal medulla of the Mammal was represented in the elasmobranch Fishes by a paired series of bodies lying along the sympathetic chain on either side, the homologue of the cortex being the unpaired interrenal body lying between the kidneys. Moore and Vincent justified this conclusion by obtaining positive chemical tests from the paired chromaphil bodies of the Elasmobranch, and Vincent (1897) showed that extracts of the paired bodies evoked the pressor response, while those of the interrenal body were ineffective. In Teleostei both the sympathetic and chromaphil elements are not so anatomically discrete as they are in the cartilaginous Fishes. Giacomini finds that the former is represented by an irregular system of nerve cells along the cardinal sinuses. The same appears to be true of *Petromyzon*. In both cases (Teleosts and Cyclostomes) the paired bodies of the Elasmobranch are represented by groups of chrome-staining cells scattered like the sympathetic ganglia in the neighbourhood of the cardinal sinuses. Gaskell (1913) obtained a pressor reaction on the cat from an extract made from tissues around these sinuses in *Petromyzon*.

The diffuse condition of the adrenaline-secreting tissue in the lower Vertebrates suggests the possibility that adrenaline may occur elsewhere in the animal kingdom; and it has already been seen that a compound allied to adrenaline is found in the salivary glands of Cephalopods. Roaf and Nierenstein (1907) noticed first that on boiling the "purple" gland of the Gasteropod *Purpura* with mordant calico a green stain indicative of the presence of an o-dihydroxyphenol was obtained. On preparing aqueous extracts of the hypobranchial organ of *Purpura*, they obtained the characteristic colour reactions of adrenaline with ferric chloride and oxidising agents. On perfusing the vessels of the frog the extracts produced a pronounced constriction; and a positive pressor response was evoked in the rabbit. Later Roaf (1911) found that certain cells of the mantle epithelium adjacent to the gland itself gave the chrome-staining reaction of Henle. Chromaphil tissue had already been described by Poll and Sommer (1903) in the chain ganglia of leeches. Their observations were confirmed and extended by those of Biedl (1910) who also recorded positive biological tests for adrenaline obtained from extracts of these cells. The search for adrenaline in Annelids has since been prosecuted more extensively by Gaskell (1914–19), who has put forward some suggestive speculations as to the phylogeny of the adrenaline-secreting apparatus and the relation of the chromaphil tissues to the evolution of the sympathetic nervous system.

A representative selection of the annelid fauna of Naples were studied by Gaskell with a view to detecting the presence of cells giving the characteristic chrome-staining reaction. Among the *Hirudinea* (leeches) chrome-staining cells were found in the ventral ganglia of all species examined. Of seventeen polychaete genera only two, *Eunice* and *Aphrodite*, were found to possess them. They were found in the Oligochaete *Lumbricus herculeus*. Wherever found they were constant in number and in position in the ganglion, six being present in three groups of two each. They have the appearance of nerve cells and Gaskell regards them as such. Extracts of the abdominal ganglia of the leech were found by Gaskell to evoke a slight inhibitory response in the virgin uterus of the cat.

Experiments carried out by Gaskell seem to indicate a nervous regulation of the automatic rhythm of the musculature of the blood vessels in those Annelids which are endowed with contractile vessels. He has also described experiments which indicate that adrenaline has an excitatory action on the pulsations of the large vessels of the leech. From the fact that chromaphil cells were only found in species with contractile vessels and never in species which do not possess a vasomotor apparatus Gaskell argues that the chromaphil cells are the cell bodies of those neurones which supply the musculature of the circulatory system in the Annelid. Direct anatomical proof of this proposition is lacking. If it is actually the case, it is plausible to suppose as he urges that (*a*) the sympathetic nervous system and the adjuvant adrenaline-secreting system are found in their earliest form in the annelid phylum, and that they consist of cells which are situated in the central nervous system that are the common ancestors of both; and that are both secretory and nervous in function; (b) that the "vascular muscle is regulated by the processes of the common ancestral cells as well as by their secretory activity."

These propositions are not however easy to place on a quantitative basis; and few zoologists would accept the conclusion that Vertebrates share with Annelids a common ancestor already equipped with a contractile vascular system. Without accepting Gaskell's hypothesis unreservedly, it does seem that a more extensive search for adrenaline in the invertebrate phyla would well repay investigation, especially as a variety of highly sensitive methods, histological, biological and chemical, are now at the disposal of any worker who elects to undertake the task.

In his speculations on the phylogeny of the adrenaline-secreting system Gaskell emphasises, it will be seen, the morphological and physiological relation between the chromaphil tissues and the sympathetic nervous system. It is interesting therefore to enquire in conclusion whether adrenaline evokes any response in the muscular tissues of animals in which no close parallel to the vertebrate sympathetic system can be recognised. As stated, Gaskell himself records acceleration of the rhythm of the lateral blood vessels of the leech by adrenaline. Elliott (1904) was unable to obtain any response in the heart of the crayfish. Carlson (1905–7) on the other hand states that adrenaline has an excitatory action both on the cardiac ganglion and the myocardium of the neurogenic heart of *Limulus*. The action of adrenaline on muscular contraction in several invertebrate preparations has been studied by Hogben and Hobson (1924) who describe pronounced increase in tone and acceleration of the cardiac rhythm of the crab, *Maia squinado*, tonic contraction of the heart muscle of the Lamellibranch, *Pecten*, the crop musculature of the Gasteropod, *Aplysia*, and of the Polychaete, *Aphrodite*, with dilutions of 1 : 50,000 or even smaller quantities. In most of these cases the action of the allied base epinine was found to be far more potent, though it is less so in relation to mammalian tissues. In the case of *Aplysia*, the crop contracting in artificial sea water was found to undergo pronounced shortening in dilutions of 1 : 5,000,000 of epinine. More recently Ten Cate (1924) has described the excitatory action of adrenaline on the hind gut of *Astacus* and the crop of *Helix*. The author (unpublished) was not able to detect any action upon the heart of the snail; but the hearts of several species of decapod Crustacea respond to adrenaline in the same way as that of *Maia*. In the case of *Maia* it was not found possible to obtain a reversible type of response, such as can be obtained from the crop of *Aplysia*. It would be interesting in the light of observations by Henze and others already mentioned to

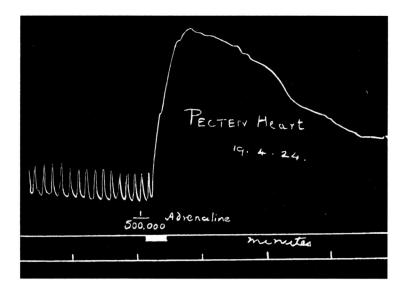

Fig. 13. Heart of pecten—Action of adrenaline.

compare the action of tyramine and adrenaline with other sympathomimetic amines on the muscular tissues of the Decapod. Brucke and Satacke (1912) record a single experiment in which an injection of adrenaline was followed by a fall of blood pressure in the lobster.

The materials investigated show such diversity from the histological and phyletic standpoint that one may question whether much importance should be attached to the action of adrenaline as an indicator of the existence of sympathetic innervation, an argument which has been widely used in physiological literature.

Since Anderson and Elliott respectively demonstrated that the efficacy of pilocarpine and adrenaline persist after degenerative section of parasympathetic and sympathetic nerves, it has only been possible to conserve the idea that such reagents act upon the motor end-organs by postulating a hypothetical myoneural junction supposed to be localised beyond the visible nerve endings.

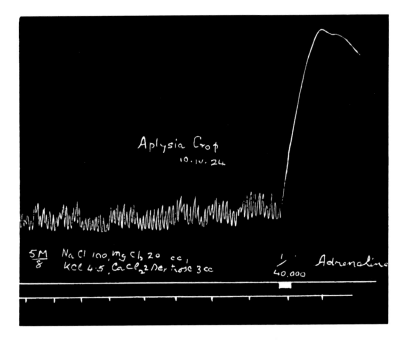

Fig. 14. Action of adrenaline on the crop of *Aplysia*.

The fact that drugs like apocodeine and atropine exclude both the action of adrenaline and pilocarpine on the one hand and the effects of stimulation of sympathetic and parasympathetic nerves on the other naturally suggests that both types of effect are referable to some common feature of the excitatory process, but in the absence of more quantitative data there seems insufficient justification for envisaging this common factor as structural in the morphological sense or as a separate chemical entity as implied by the term "receptor substance." It is even legitimate to question whether such terms as parasympathomimetic and sympathomimetic would have been employed extensively, if the recognition of the action of muscarine on the heart of *Helix* and of adrenaline on the crop of *Aplysia* or the heart of *Limulus* had preceded that of the action of pilocarpine on the heart of the frog and of adrenaline on the ileum of the cat.

REFERENCES

GASKELL. *Phil. Trans. Roy. Soc.* 205, 1914; *Journ. Gen. Physiol.* 2, 1919.

HOGBEN and HOBSON. *Brit. Journ. Exp. Biol.* 1, 1924.

ROAF. *Quart. Journ. Exp. Physiol.* 4, 1911.

ROAF and NIERENSTEIN. *Journ. Physiol.* 36, 1907.

SWALE VINCENT. *Proc. Roy. Soc.* 61, 62, 1897.

TEN CATE. *Arch. Neerland. Physiol.* 9, 1924.

Chapter III

INTERNAL SECRETION AND THE CHROMATIC FUNCTION

§ I

The physiological significance of adrenaline has been dealt with first because of the possibility of its widespread occurrence in the animal kingdom. In what follows we shall confine our attention to the relation of internal secretion to definite manifestations of vital activity or chemical processes within organisms, beginning with the chromatic function. Colour response is seen in representatives of both vertebrate and invertebrate phyla; but from the present standpoint the former alone need concern us.

Colour change in the chameleon was known to writers of antiquity. Aristotle wrote concerning it: "the change in colour of its skin takes place when it is filled with air. It can acquire either a black colour like that of the crocodile or ochreous like that of the lizard, or spotted with black like that of the panther: and this change takes place over the whole body; for the eyes also change like the rest of the body, and so doth the tail." Towards the end of the seventeenth century Perrault presented a thesis on the same subject to the Paris academy; and since that time it has repeatedly attracted the attention of naturalists without noteworthy advance in our knowledge of its intimate nature, until Milne Edwards (1848) recognised that pigmentary response in the Chameleon is brought about not through any chemical changes in the skin pigments, but by the mechanical distribution of the dark colouring matter contained in deep seated receptacles whose processes ramified among the yellow pigment immediately below the epidermis. Brücke (1854) first demonstrated the cellular character of the elements associated with colour changes in the chameleon, and gave the first satisfactory account of their anatomical basis in the behaviour of the pigmentary effector organs in this reptile.

Among the first to investigate the chromatic function from the physiological standpoint was Lister, who in his memoir on colour response in the frog wrote:

The pigmentary system also promises to render good service in toxicological enquiry. Hitherto in experiments performed on animals with that object, attention has been directed chiefly, if not exclusively, to the effects produced upon the actions of the nervous centres, the nerves, and muscles. In the pigment cells we have a form of tissue with entirely new functions, which, though apparently allied to the most recondite processes, yet produce very obvious effects....Such experiments are so readily performed, and the effects produced are so obviously indicated by changes in the colour of the integument, that I venture to recommend this method of investigation to those who are occupied in studying the action of poisons.

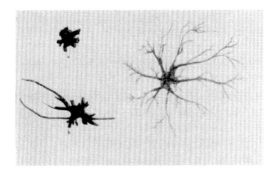

Fig. 15. Three stages in melanophore expansion in the Lizard *Tarentola* (Schmidt).

§ II

Such information as we possess with regard to the possible co-operation of endocrine factors in reptilian colour response is derived from a study of the phenomena which occur in the so-called "horned toad," *Phrynosoma*, the only reptile which has been made the subject of recent experimental treatment through the researches of Parker (1906) and Redfield (1918).

In the Horned Toad bodily colour varies between a fuscous shade associated with the outward migration of the melanophore granules into the processes which ramify among the yellow interference cells immediately below the epidermis, and a pale cinnamon buff tint resulting from concentration of the pigment granules in the cell body of the melanophores. When kept on a neutral background there is a daily rhythm of colour change in Phrynosoma. At night the melanophores are contracted, giving the skin the appearance of pallor. In the early morning the skin becomes uniformly dark through "expansion" of the melanophores. But during the heat of the day—in its warm natural surroundings—the melanophore pigment contracts, and as evening approaches a second expansion supervenes, until night descends. The condition of pallor in natural surroundings is, therefore, seen at night and at midday. In the cooler parts of the day the skin is fuscous.

The exact part which light and temperature respectively play in promoting this sequence of reactions, has been made the subject of investigation by Parker (1906) and Redfield (1918). From their work it appears that the melanophores of this lizard respond to light and darkness, warmth and cold, in the manner generally characteristic of Reptiles, as indicated above, i.e., bright illumination and low temperature promote darkening of the skin, while warmth and darkness bring about pallor. Light and heat interact so that the effect of the former is only significant within a comparatively restricted range. It is thus that, in natural conditions, living, as these creatures do, in a warm climate pallor intervenes during that part of the day, when the temperature rises to a maximum.

But in addition to this response to direct illumination, the Horned Toad reacts in bodily coloration to the character of the substratum and to mechanical irritation or disturbance. Any nocuous stimulus, such as electrical excitation of the roof of the mouth or the cloaca, evokes pallor in fuscous individuals within a few minutes. Kept for several days on a white background the animals remain fuscous in bright light.

Redfield has recorded the results of careful experiments carried out with a view to locating the receptors and co-ordination mechanisms involved in these modes of response. He finds that local exclusion of light from, and application of heat, restricted areas of the skin produce a local contraction of the melanophores in the region to which the stimuli are applied, without affecting the colour of the skin in other parts of the body. Furthermore, local illumination produces a local expansion of the melanophores without affecting the pigmentary effectors of other regions, while local reduction of temperature maintains locally, though it apparently cannot initiate, a state of melanophore expansion already established. These experiments admit the possibility that the response to heat and to light, in the case of animals kept on a neutral background, is propriogenic in character and results from the direct reactivity of the pigmentary effector organs to incident stimuli: it is not conceivable that these results could be brought about by hormonic regulation through the circulatory system. Redfield states that such local responses can be evoked after the entire nerve supply of the affected region has been severed. If this is so, there would seem to be no alternative to accepting his conclusion that light and temperature can act directly upon the melanophores, without the intervention of either a freely circulating hormone or a nicely adjusted system of reflex arcs.

Nevertheless, Redfield is driven to the conclusion that there is, superimposed on this primary reactivity of the melanophores of the Horned Toad to light and heat, a co-ordinating mechanism which will account for the generalised condition of pallor following "excitement," and the peculiar modification of the normal reaction to light in virtue of the background upon which the animal is kept. For the latter response the appropriate receptor is the eye; since blinded individuals no longer display the "adaptive" response to the brightness or darkness of the substratum. If Horned Toads, which have been kept upon a background of dark cinders, are transferred to one of white sand, they become noticeably paler after one day, and reach within five days a condition of maximum pallor. When however the eyes are blindfolded, exposure to the same surroundings for several weeks does not result in the disappearance of the dark condition; and the results of carefully controlled experiments showed that this is not due to the mechanical influence of the bandage, but can only be interpreted on the assumption that response to the nature of the background arises through stimuli received in the first place through the organs of vision. In exploring the possibilities of pigmentary control through the eyes, it is of interest to contrast the relatively slow and accumulative nature of the response to background with the more rapid primary reaction of incident light. We shall consider the mechanism of pigmentary response to background in Amphibia later on, where it will be seen that pituitary secretion appears to be the significant factor. In the case of the horned toad, the intimate nature of the background response was not investigated by Redfield, who devoted his investigation of the co-ordinating mechanism in Reptilian colour response to the peculiarly characteristic phenomenon of excitement pallor, and by an ingenious and painstaking series of experiments has arrived at the conclusion that here too internal secretion plays an important part in the process.

The method adopted to induce pallor by nocuous stimulation was the application of a faradic current to excitable areas, such as the cloaca or roof of the mouth, which, as we have seen, results in general contraction of the melanophores throughout the skin of the whole body. That such treatment results in uniform melanophore contraction in dark animals even after the denervation of definite areas of the skin, such as can be achieved by severing all the nervous connections of a limb, suggests at once that the melanophores are susceptible to stimuli received through the circulatory system. Several lines of experimental evidence converge to this conclusion: in particular, the possibility of evoking pallor by transfusion of blood from an excited animal. It has been known for some time that extracts of the suprarenal medulla induce melanophore contraction in Fishes and Amphibia. In the Horned Toad destruction of the cord between the eighth and thirteenth vertebrae prevents pallor after faradic stimulation of the mouth. In these circumstances the body cavity may be opened without producing melanophore contraction; and when the adrenals of such lizards were stimulated, contraction of the melanophores occurred throughout the entire body after the lapse of only a few minutes. It did not occur, however, in the hind limb after ligature of its arterial supply. When the ligature was removed on the other hand, the skin of the leg rapidly assumed the condition of extreme pallor. Collateral evidence in favour of the possibility that adrenal secretion determines excitement pallor in Reptiles was provided by analysis of the blood sugar content. Redfield found that the blood sugar content was significantly higher in lizards after the production of pallor by nocuous stimulation. To sum up the evidence from these and other experiments briefly, it seems clear: (1) that adrenaline provokes the contraction of reptilian as well as amphibian melanophores; (2) that the adrenal glands of Reptiles contain a substance which has the same action as adrenal extracts obtained from mammalian glands; (3) that there is indirect evidence that adrenal activity is associated with "excitement" in Reptiles and (4) that removal of the adrenals in Phrynosoma (Horned Toad) in most cases prevented melanophore contraction to nocuous stimulation.

Redfield does not regard the adrenal-secreting mechanism as the only co-ordinating factor in the production of pallor after nocuous stimulation, and advocates the view that nervous impulses also take part in the process. On this point his interpretation seems to the present writer to be open to criticism. Few investigators who have studied the physiology of colour response have paid sufficient attention to an important consideration, upon which Langley has rightly insisted in his recent monograph on the Autonomic system. In view of what is known concerning the effect of anaemia upon melanophores in Reptiles and Amphibia, and more especially the evidence of endocrine control in colour response, it is evident that any experiments designed to show the effects of nervous stimulation must be carried out in such a manner as to exclude the possibility that the phenomena observed do not arise secondarily through interference with the blood supply.

In these experiments Redfield used animals in which the anterior part of the spinal cord had been destroyed, so that, possibly through destruction of the tracts involved in reflex activation of adrenal secretion, the animals no longer responded to nocuous stimuli by pallor. In no case did he obtain local melanophore reactions by stimulation of the spinal nerves in these specimens. The results of sciatic stimulation were very variable. And in no instance did animals with both adrenals and the central nervous system intact fail to give the characteristic and synchronous colour change following nocuous stimulation of the mouth, when the nerve supply to different regions of the skin had been destroyed. For evidence of nervous control Redfield relies on the fact that in certain specimens removal of the adrenals did not abolish excitement pallor, and in these cases stimulation of the mouth after spinal transection produced pallor anterior to the region of transection. For a correct interpretation of these data, there are, however, two considerations which should be taken into account. First, with regard to the failure of epinephrectomy to abolish excitement pallor in the individuals mentioned, it must be remembered that accessory chromaphil tissue is commonly met with in the lower Vertebrata; secondly, we require more first-hand knowledge of the effects of nervous transection and stimulation on the arteriole and capillary supply of the regions affected, before accepting the melanophore response to nerve stimulation as a direct one. On this point the evidence is inadequate, and Redfield observes that "the responses of melanophores to direct stimulation and to hormones evidently suffice to bring about all ordinary melanophore reactions, without the aid of nerves which connect with these cells directly."

It may be stated that there exist, at present, no well-authenticated experiments demonstrating positive effects of nerve section or stimulation on pigmentary response in Reptiles or Amphibia conducted in such a way as to exclude the possible influence of concomitant vasomotor changes. This being so, it is hardly justifiable to postulate a direct innervation of reptilian melanophores, until there is actual histological evidence that they receive nerve fibres. Such evidence is not yet available.

§ III

Perhaps the most conclusive evidence of internal secretion in the field of comparative physiology is seen in the control of colour response in Amphibia. In Amphibia the cutaneous pigmentary effectors are, as far as is known, restricted to three types which are practically ubiquitous. These are (a) epidermal melanophores; (b) dermal melanophores; and (c) dermal xantholeucophores or interference cells. In some Amphibia, the xantholeucophores are said to be non-contractile in adult life. In others, however, they respond by contraction when the melanophores expand and *vice versa*. It is also a fact that under ordinary external conditions the melanophores are permanently expanded in adult life in many melanic species, e.g. old black axolotls and amblystomas. In addition to the foregoing there are other pigmentary effectors of more deep-seated situation. These are (*a*) the retinal pigment cells; and (*b*) the "internal" melanophores present in many connective tissues, lining the peritoneum and lymph spaces. Except when otherwise stated subsequent remarks apply only to the cutaneous, more especially the dermal (or corial) melanophores, whose activity constitutes the dominant factor in bodily colour changes in Amphibia.

The exact manner in which the melanophores "contract" and "expand" has been subject to much controversy. Some workers following v. Wittich (1854), Nimmermann (1878), Carnot (1896), hold that the melanophores are amoeboid, contracting bodily. This view has been advocated recently by Hooker (1914). Others following Lister (1838), Müller (1860), Kahn and Lieben (1907), Dawson (1920) affirm that the cell processes are not withdrawn in the "contracted" phase which results purely from a migration of pigment granules from the cell processes into the central mass. The latter view claims more support from recent investigators.

One of the most important factors involved in colour response in adult Amphibia is moisture. This, of course, does not enter into the responses of larval and aquatic forms as a significant element in the normal rhythm of colour change. It seems to be generally agreed that humidity promotes darkening of the skin in adult Anura, and my own observations indicate that, in nature, moisture is the predominant factor which determines colour change, since except at low temperatures light and darkness have no effect on frogs placed in dry surroundings. With the exception of Dawson's observations on *Necturus* (1920) it is universally affirmed that warmth promotes the contracted phase and cold tends to induce darkening of the skin in adult Anura (v. Wittich, Biedermann, Lieben, Hargitt, *et al.*), larval Anura (Cole 1922), and adult Urodela (Flemming, Laurens)[1]. Another factor worthy of mention is the oxygen content of the surrounding atmosphere. It is generally agreed that oxygen deficiency induces darkening, and the author's observations on the frog agree in this respect with that of others. For all adult Amphibia, with the exception of *Necturus* (Dawson),

---

[1] The Anura (tailless forms) include frogs and toads: the Urodela (caudate amphibians) include the newts and salamanders.

light, if it has any effect, promotes the contracted phase, darkness or dull background the expanded. Probably the distinction between background and shade has been greatly overstressed. This mode of response appears to be under visual control both in adult (Lister, Rogers) and larval (Babak, Laurens) Amphibia. Anuran tadpoles with the exception of *Rana pipiens* (Hooker) react in the same way as the adult by contraction in strong light and expansion in darkness. In Urodele larvae (Laurens) there is a primary response to light by expansion and to darkness by contraction, followed after prolonged exposure by the more usual reaction of pallor in the presence and darkening in the absence of light. It is this secondary response which is abolished by the removal of the eye and section of the optic nerve. Observations on the adult frog are in agreement with those of Rogers (1906) on the adult salamander, namely, that in the presence of moisture, within the limiting range of temperature for which light is a significant factor, the dermal melanophores of normal frogs contract in bright light, whereas those of the blinded frog remain permanently expanded in light and shade alike. Adaptation to background is thus under control of the organs of vision in Amphibia as in Fishes.

In order to make clear the precautions necessary in controlling experiments on colour response, the following table sums up the normal pigment responses of the common frog, *R. temporaria.*

Background	20° C.	10° C.
Light background		
(a) Dry	Pallor	Generally pale
(b) Moist	Pallor (Epidermal melanophores expanded)	Darkening
Shade or dark background		
(a) Dry	Pallor	Partial darkening (Epidermal melanophores contract)
(b) Moist	Darkening	Darkening
Darkness		
(a) Dry	Pallor	Partial darkening
(b) Moist	Darkening	Darkening

Note:—"Pallor" implies contraction, "darkening" expansion of both dermal and epidermal melanophores.

The extremely protracted latent period, which intervenes between the application of the appropriate stimuli and the production of a particular colour change in consequence, has been emphasised by few investigators. Had more attention been paid to this fact, it can hardly be doubted that the search for an alternative to nervous control would have been made at an earlier stage. Darkening and pallor in response to the optimum conditions defined above is never complete in frogs before at least one and a half hours and more usually six to twenty-four hours have elapsed. The only exact observations recorded in this connexion are those of Laurens (1915) on Urodele larvae. The following citation is taken from Laurens' paper on *Amblystoma* larvae:

1. Expansion of the melanophores of seeing larvae in the light one and a half to two hours.
2. Expansion of the melanophores of eyeless larvae in the light two to three hours, and contraction of the melanophores of seeing larvae in darkness two to three hours.
3. Contraction of the melanophores of eyeless larvae in darkness four to five hours.
4. "Secondary" contraction of the melanophores of seeing larvae in the light three to five days, and "secondary" expansion of the melanophores of seeing larvae in darkness five days or more.

Laurens does not give the temperature conditions to which such periods are subject.

The synchronous character of the colour responses which occur in Amphibia implies that a regulatory mechanism of some kind controls the pigmentary reactions. Early work on this problem was directed exclusively to the possibility that this regulatory mechanism is provided by the nervous system. There is, however, no unanimity regarding the effects of nervous section and stimulation, spinal transection, and extirpation of the sympathetic ganglia. Shortly after the publication of Oliver and Schafer's researches, Corona and Moroni (1898) noticed the effect of adrenaline in producing melanophore contraction in frogs, an observation confirmed by the researches of Lieben (1906). A new line of attack on the problem has emerged from the study of developmental physiology during the last decade; and to the discussion of this the remainder of this chapter will be devoted.

The rôle of the pituitary in the regulation of pigmentary changes in Amphibia first attracted attention when Smith (1916) and Allen (1917) independently developed a technique for the ablation of the hypophysis anlage in anuran embryos, at the time of its appearance as a wedge-shaped invagination of cells from the oral ectoderm. They both found that the tadpoles a few days after the operation lose their normal dark coloration and assume a silvery white appearance on account of which they were originally described as albinos. The precise significance of this change was not however appreciated immediately by these investigators. Smith considered that the silvery appearance of the hypophysectomised larvae was due primarily to a reduction in the number and melanin content of the melanophores and to the persistent expansion of the xantholeucophores. Allen interpreted the albino effect as a consequence of migration of the epidermal pigment cells to deeper positions in the body; but maintained correctly that all the melanophores of albino larvae were in the contracted state. Atwell (1919) placed such "albinos" in dilute extracts of bovine pars intermedia and observed that they then assumed the black coloration of normal tadpoles. Swingle (1921) investigated the effects of implantation of the intermediate lobe of adult frogs of three species (R. catesbiana, R. pipiens, R. clamata) into light-coloured larvae, and described a darkening of the skin to the expansion of both dermal and epidermal melanophores and contraction of the xantholeucophores. About the same time Smith (1921) reinvestigated the effects of hypophysectomy on the colour reactions of tadpoles and confirmed Atwell's observations but controverted the interpretation which that writer placed on them, reaffirming the conclusion that the expansion of the xantholeucophores and actual reduction in the amount of melanin in the skin were the main factors contributing to the "albino" effect. At the same time Smith proved by reciprocal skin changes that the reaction of the pigmentary effectors of all three kinds is due not to nervous agencies but to the character of the tissue fluids in which they are bathed. Smith however emphatically asserts that "the deep melanophores of the albino and normal larvae react identically under all tests tried by the author."

Simultaneously a series of researches on the chromatic function in the common frog was begun by Hogben and Winton (1922) with a view to harmonising these conflicting observations. The

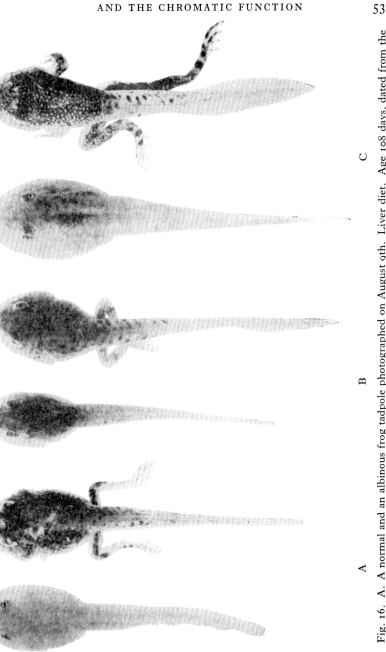

A

B

C

Fig. 16. A. A normal and an albinous frog tadpole photographed on August 9th. Liver diet. Age 108 days, dated from the operative stage. × 2½. B. A normal and an albinous frog tadpole of the same age and photographed on the same date as the specimens shown in figure A. Posterior-lobe diet. Age 108 days. × 2½. C. A normal and an albinous frog tadpole of the same age and photographed on the same date as the specimens shown in figures A and B. Anterior-lobe diet. Age 108 days. × 2½.

results of preliminary experiments may be briefly summarised under four headings:

(1) The pituitary (pars intermedia and nervosa) of Mammals, Birds, Amphibia and Fishes, contains a specific stimulant capable of inducing contracted melanophores of adult and larval Amphibia (Anura and Urodela) to undergo maximum expansion.

(2) This property is not shared by such drugs (e.g. histamine) as simulate the physiological action of pituitary extracts in other

Fig. 17. Two frogs. Right, individual injected six hours previously
with pituitary extract from a foetal ox: left, control.

respects, nor is it shared by other tissue extracts examined, namely, spleen, brain, testis, ovary, pancreas, liver, muscle, adrenal, pineal and salivary gland.

(3) The melanophore response is a very sensitive indicator of pituitary extracts. The gland of a single frog contains sufficient to induce darkening in some fifty or more individuals of the same species.

(4) The action of the melanophore stimulant in pituitary extract is direct and local, independent of concomitant vasomotor effects.

Krogh in the course of his experiments on perfusion of the vessels of the frog found that pituitary extracts will produce darkening at a dilution of one in ten million. Fenn (1924) obtained melanophore expansion by the same method at a dilution of 10^{-10} fresh gland substance per c.c. solution.

Later the effects of removal of the whole pituitary, controlled by comparison with effects of exposure of the brain, section of optic nerves and removal of the anterior lobe alone were investigated by Hogben and Winton (1923) in the same species, and analogous experiments to those of Smith and Allen on the axolotl

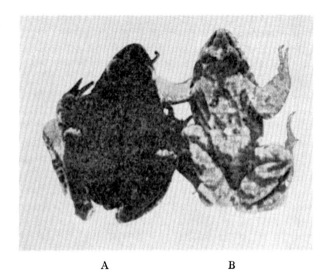

A   B

Fig. 18. Two frogs, 19 days after operation, kept in wet shady conditions at 10° C. A. Partially hypophysectomised (removal of anterior lobe only). B. Complete hypophysectomy. (Photographed by Mr J. Chisholm, artist to the Animal Breeding Research Department.)

larva of the Mexican salamander were subsequently carried out by Hogben (1924) whose observations have since been confirmed by those of Swingle. The outstanding results of these researches may be epitomised as follows:

(1) After removal of the whole pituitary in adult frogs, as in axolotls and anuran tadpoles (Smith, Allen and Atwell), the individuals so treated remain permanently pale with the melanophores in maximum contraction, although subjected to conditions which inevitably induce darkening of the skin in normal animals.

(2) Melanophore expansion follows pituitary injection in hypophysectomised Amphibia, but hypophysectomised individuals so treated regain their characteristic pallor, although subject to conditions which inevitably evoke melanophore expansion in the normal animal.

(3) A comparison of the minimal standardised dose of a sample of pituitary extract requisite to induce melanophore expansion in normal and hypophysectomised frogs, under conditions in which the intensity of external factors tending to promote pallor were varied, favours the view that melanophore contraction and expansion in the intact animal is correlated with a fluctuating amount of pituitary secretion in the circulation. (See accompanying table.)

Action of Post-Pituitary Extract on the Hypophysectomised Frog

Normal series			Hypophysectomised series		
Dose in cc.	Weight in grms.	Condition of melanophores 2 hours later	Dose in cc.	Weight in grms.	Condition of melanophores 2 hours later
(1) 22° C.—					
0·00001	17	Contracted	0·00001	18	Contracted
0·00005	17	,,	...	...	...
0·0001	22	,,	0·0001	18	Contracted
0·0005	22	Reticulate	0·0005	18	Reticulate
0·001	24	,,	0·001	20	Stellate
0·01	25	,,	0·01	21	Reticulate
0·1	...	...	0·1	24	,,
(2) 12° C.—					
0·000001	18	Contracted	0·000001	17	Contracted
0·000005	22	,,	0·000005	17	,,
0·00001	23	,,	0·00001	21	,,
0·00005	23	Stellate	0·00005	18	,,
0·0001	24	,,	0·0001	22	Stellate
0·0005	26	Reticulate	0·0005	23	Contracted
0·001	26	,,	0·001	24	Reticulate
0·005	33	,,	0·005	24	,,
0·01	32	,,	0·01	24	,,
0·1	30	,,	0·1	26	,,

(Hogben and Winton.)

The regulation of colour response by fluctuating pituitary secretion is thus probably adequate to interpret all the accredited

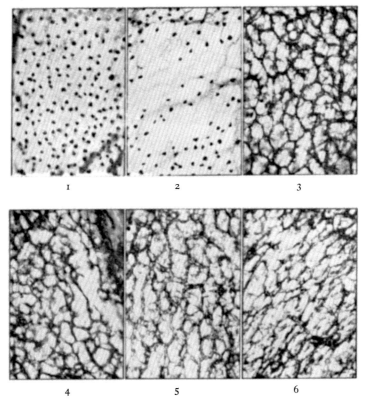

Fig. 19. Microphotographs of melanophores in web of the foot of the common
frog (*Rana temporaria*). 1. Normal pale animal (dry, brightly illuminated at
20° C.). 2. Hypophysectomised (both lobes), 19 days after operation (wet,
shady, 10° C.). 3. Hypophysectomised (both lobes), after injection of posterior
lobe (ox pituitary) extract, six hours later. 4. Normal dark animal (wet, shady,
10° C.). 5. Partially hypophysectomised (anterior lobe only), 19 days after
operation (same conditions as 4). 6. Exposure of brain 19 days previously
(same conditions as 4 and 5).

phenomena in adult Amphibia, without invoking a direct innervation of melanophores. We may justifiably conclude the fact that
in Urodela as in Anura pituitary secretion is controlled by various
(e.g. thermic) receptors in the skin, and is reflexly inhibited by

light acting on the retina. This fully explains why in the sala-mander, *Diemyctilus* (Rogers), within the range of external con-ditions for which light is the significant factor, section of the optic nerve was found to result in permanent melanophore expansion, although transection of the cord was without effect on the rhythm of colour response.

Fig. 20. Two hypophysectomised axolotls one month after operation. Right individual showing pallor consequent upon pituitary removal. Left individual in which normal colour has been restored by injection of pituitary extract.

The possible relation of the melanophore stimulant to the other pituitary autacoids will be discussed elsewhere. In most respects it has the same general properties. It is soluble in water, with-stands boiling, is insoluble in absolute alcohol, ether and acetone, is destroyed by trypsin and by acid hydrolysis. For quantitative experimentation the only methods of assay are based upon deter-mination of the minimal dose, either as originally proposed by

Hogben and Winton (1922) on parallel series of frogs of corresponding size kept under standard conditions to promote pallor, or as recently proposed by Fenn (1924) by perfusion of the vessels of a single animal. A tolerable measure of precision (20 per cent.) can be attained by either method in experienced hands but neither can be said to be very satisfactory for accurate work.

Contraction of melanophores in Amphibia as in Reptiles and Fishes is brought about by injection of adrenaline; but this cannot be accepted as proof of sympathetic innervation in the absence of direct evidence. In anuran tadpoles injection of extracts or feeding with substance of pineal glands produces an extreme condition of pallor (McCord and Allen). But this reaction does not occur in the adult nor in larval Urodela; and Laurens found that removal of the pineal in *Amblystoma* larvae had no appreciable effect on the chromatic function.

As stated earlier previous attempts to interpret the co-ordinating mechanism in colour response among Amphibia have been almost exclusively concerned with exploring the possibility that such a mechanism is nervous. The methods adopted include peripheral and central nervous stimulation and section, on the one hand, and a comparative study of the pigmentary reactions following administration of drugs with characteristic physiological action on the other. Apart from Langley (1921) few writers seem to have clearly appreciated the necessity of discriminating between effects following experimental treatment due to direct innervation of the pigmentary effector system, as opposed to indirect consequences of interference with the blood supply. Now we know that there exists an endocrine regulation of colour response acting through the circulation in Amphibia: the consideration which Langley emphasises becomes even more significant, and the more so in the light of recent work by Krogh (1920–22) and his colleagues upon nervous control in capillary phenomena. Spaeth, than whom no writer has insisted more strongly upon the nervous control of colour response in Fishes, admits that there exist "no satisfactory demonstrations of the nerve endings in the melanophores of Reptiles and Amphibians."

Briefly the main issues raised in connexion with the alleged nervous control of amphibian colour response may be summarised thus:

(1) The current doctrine that melanophores are innervated in Amphibia rests on neither direct histological nor unequivocal experimental data.

(2) The problem has been investigated in the past by methods which include peripheral and central stimulation and transection and the colour responses following drug administration. Such procedure fails to discriminate between direct nervous control of the melanophores and concomitant vasomotor consequences of experimental procedure.

(3) In no case have such experiments been carried out with rigid controls under optimum external conditions acting in the opposite sense to the recorded effects; in no case have photographs of these effects been published for the benefit of other investigators, and on no relevant points relating to nerve section and stimulation is there unanimity in the testimony of different observers.

(4) Judging from the effect of drugs on the hypophysectomised frog we can legitimately conclude that, even if a nervous mechanism for regulating colour control exists, it plays no significant rôle in the rhythm of normal colour response in Amphibia.

(5) The hypothesis of pituitary secretion fluctuating in correspondence with the action of natural stimuli tending to promote colour response is in the existing state of knowledge adequate, at least in adult Amphibia, to interpret all the salient facts.

This hypothesis of colour response is not intended to give an account of pigmentary changes in Reptiles and Fishes, where there are indications that the pigmentary effector organs respond both to external conditions and to internal secretions in a somewhat different manner. One may urge, however, that the rôle of internal secretion in regulating colour response in other Vertebrata might profitably be made the subject of more extensive research. In Fishes there is no definite histological evidence of the innervation of the pigmentary effectors (Ballowitz) and unequivocal experimental evidence (Pouchet, v. Frisch, Wyman) that direct nervous control is an important factor in their colour responses. The observations of Houssay (1924) who has recently repeated and confirmed those of Hogben and Winton on adult Anura show that pituitary extract does not produce expansion of melanophores either in Reptiles or in Fishes. In the scales of the Atlantic minnow

(*Fundulus*) Spaeth (1917) has shown that the melanophores contract in response to exceedingly minute quantities of pituitary extract. As far as the available evidence permits any inference it would seem that the predominant factors in the regulation of colour response in Reptiles, Amphibia and Fishes are different. The nature of the co-ordinating mechanism controlling the pigmentary effector system of Crustacea is unknown. Keeble and Gamble do not regard it as nervous; and the further investigation of colour response in Crustacea might possibly lead to an extension of our knowledge of internal secretion to the invertebrate phyla.

REFERENCES

For complete bibliography see:

FENN. *Journ. Physiol.* 59, 1924.
HOGBEN. *The Pigmentary Effector System.* Oliver and Boyd, Edinburgh, 1924.
HOUSSAY and UNGAR. *Revista Assoc. Med. Argentin.* 37, 1924.

Chapter IV

ENDOCRINE FACTORS IN SECRETORY PROCESSES

§ I

In this chapter we shall deal with the part played by endocrine factors in regulating secretory processes in the body. It was in relation to the regulation of pancreatic secretion that the phenomenon of chemical co-ordination was first clearly established through the investigations of Bayliss and Starling in the opening years of the present century. It had long been recognised on the basis of researches by Claude Bernard, Heidenhein, Pawlow and others that the secretory activity of the pancreas is normally called into play by events occurring in the alimentary canal. Pawlow and his pupils in particular had shown that the flow of pancreatic juice is normally initiated by the entry of the chyme into the small intestine; and could be induced by introducing acid fluid into the duodenum. Later Popielski (1900) and Wertheimer and Lepage (1901) showed that this response occurs even after section of the vagi and splanchnics with complete extirpation of the solar plexus. Bayliss and Starling (1902) confirmed and extended such observations, succeeding in provoking pancreatic secretion after section of vagi and splanchnics in an animal whose spinal cord was destroyed from below the sixth thoracic vertebra. They then separated a loop of intestine (jejunum) from all its nervous connexions, and demonstrated the efficacy of acid when introduced into the enervated loop to call forth a copious flow of pancreatic juice.

These experiments clearly demonstrated that introduction of acid into the intestine could evoke pancreatic secretion without the intervention of any nervous connexions between the pancreas and the gut. The secretion was clearly not reflex. It might on the other hand be due to direct excitation of the pancreatic cells by a substance or substances conveyed from the intestine to the gland through its intact blood supply. Wertheimer and Lepage had shown however that injection of acid itself into the circulation does not activate the pancreas: the exciting agent could not therefore be the acid itself. There remained the possibility that the action of the acid on the cells lining the gut liberated into the blood stream a substance capable of exciting the pancreas to active secretion.

This surmise was put to experimental test and verified by Bayliss and Starling (1902). The mucous membrane scraped off a strip of intestine was ground with sand, boiled in 0·4 per cent. HCl, and the filtered extract injected into the circulation of a dog. The result was an immediate flow of juice from the pancreatic duct beginning about a minute after the injection. The neutral extract was without action. To the active substance produced by the action of acid the term *secretin* was given, its precursor being designated prosecretin.

These observations were abundantly confirmed in the same year by the work of Camus and Gley, Brissow and Walter, Delezenne, Stassano and Billon and Wertheimer. To leave no alternative to the interpretation which they advocated Bayliss and Starling themselves investigated the fall of blood pressure accompanying pancreatic activity in consequence of the injection of duodenal extract; and showed that the response could be brought about equally well with a preparation from which the depressor substance had been removed by alcoholic extraction. It could not therefore be argued that the flow of pancreatic juice was a secondary consequence of vasomotor disturbances. Furthermore, they showed that secretin was a specific product of the cells of the intestinal mucosa, since it was not possible to make extracts with similar properties from other tissues of the body. Later (1903) they demonstrated the production of pancreatic activity by extracts prepared from the intestinal mucosa of the tortoise, frog, salmon, dogfish and skate, showing at the same time that secretin activates the pancreas of the monkey, rabbit, cat and goose. Apparently therefore an identical endocrine mechanism exists throughout the vertebrate series.

Later researches have added surprisingly little to the account given in the original paper by Bayliss and Starling, a memoir that should be read by every student of endocrinology as a model of discriminating treatment in experimental work on such lines. The significance of the fall in blood pressure which accompanies the injection of secretin prepared in the manner described above has recently been investigated by Parsons (1925) whose observations are not identical with those of Bayliss and Starling though confirmatory on the main issue, i.e. the independence of the vasomotor and secretogogue effects. Parsons finds that there is no change in the blood pressure when 0·4 per cent. HCl is introduced into the intestine: the acid acts as a secretogogue without any lowering of the blood pressure. The isolation of β-iminazolyl-ethylamine in intestinal mucous membrane by Barger and Dale (1911) and the presence of this powerful depressor base in many tissue extracts according to Abel and Kubota (1919) suggests that the vasodilator substance may be histamine. But the estimations of Parsons shows that the quantity of histamine in the preparations of secretin which she employed in her investigation were insufficient to account for their depressor activity. This however does not justify the conclusion that secretin itself has a vasodilator action: there are undoubtedly other vasodilator substances besides histamine in tissue extracts. Whatever the nature of the depressor base there is no doubt that the secretogogue and depressor properties of secretin preparations are physiologically independent (*vide infra*).

Of the chemical properties of secretin Bayliss and Starling themselves showed that it withstands boiling in acid medium, but is easily destroyed by pancreatic juice, oxidising reagents and many metallic salts; that it is not precipitated from its solutions by tannic acid, alcohol or ether; and that it is slightly diffusible through a parchment membrane. From its precursor it may be extracted not only with dilute acids but with 70 per cent. alcohol and strong soap solutions (Stepp). Secretin has not yet been isolated. Its insolubility in lipoid solvents suggest that it is not a lipoid. Its destruction by trypsin must not be taken as proof that it is of a polypeptide-like nature. This property is shared by insulin and by the pituitary autacoids. Tryptic preparations free from esterases have not been employed; and as with the other autacoids destroyed by trypsin the action of the enzyme may be exerted on some substratum to which these by no means stable substances are absorbed. A method of obtaining a highly concentrated extract has been described by Dale and Laidlaw (1912) who proceed as follows. A 50 per cent. decoction of fresh mucosa in water is boiled and filtered. The coagulum containing practically all the secretin is squeezed dry. It is then made up to 25 per cent. fresh tissue substance in a mixture of acetic acid 2 per cent. and mercuric chloride 1 per cent. This is boiled and filtered. Caustic soda is added till the solution is not quite neutral; and the precipitate so formed is removed and suspended in water. This mercuric compound is decomposed for removal of the mercury either by means of sulphuretted hydrogen or 75 per cent. alcohol. Its further concentration can be effected by precipitating from the solution an amorphous picrate which can be redissolved in a solution of sodium carbonate.

Since the preceding paragraphs were written, J. Mellanby (1926) has announced the discovery of a method isolating secretin, by extracting with alcohol, and absorbing the active substance with the aid of bile salts. In this preliminary communication reasons are given for supposing that secretin is a polypeptid and the important fact is stated that while having a powerful action on the external secretion of the pancreas, the purified substance has no depressant action on the blood pressure.

Apart from the chemical nature of secretin an important issue still remains for further investigation. The experiments of Bayliss and Starling proved once and for all that there exists a mechanism capable of co-ordinating the activity of the pancreas with the entry of food into the bowel. And this discovery, to quote Swale Vincent, involved "an important modification of our conception as to the empire of the nervous system." But pancreatic secretion can be induced by nervous stimulation and we have still to enquire to what extent the production of secretin can be properly regarded as the effective agent in normal regulation of pancreatic secretion. The question is not fully decided, but the somewhat different quality of the juice which is secreted under the action of secretin and nervous stimulation respectively suggests that both agencies in the intact animal co-operate in controlling the secretory activity of the gland.

It is not without value as illustrating the difficulties that attend a critical investigation of these problems to turn from the part played by secretin in the activation of the pancreas to an examination of the evidence for the view that reflex gastric secretion is augmented by an analogous mechanism. Edkins (1906) found that watery extracts prepared from the pyloric mucosa by boiling, when injected into the circulation of the dog evoke a flow of gastric juice. Extracts made in cold water, peptone, glucose and glycerine also contained this substance. It was not however found possible to extract it from the mucous membrane of the fundus. It therefore seemed not improbable that the first products of digestion act on the pyloric mucosa thereby liberating a hormone carried by the blood stream to all the glands of the stomach.

This conclusion has been adversely criticised on various grounds, though the substantial correctness of Edkins' observations is not questioned. Popielski (1909) showed that a number of compounds, small doses of Witte's peptone, extracts of other organs, etc., when injected into dogs with a gastric fistula provoke the secretion of gastric juice; and all such substances lower the blood pressure. In addition such states as anaphylactic shock, blood transfusion, morphine narcosis, which are also associated with lowered blood pressure are accompanied by increased gastric secretion. Lim and his co-workers (1923–5) who have investigated the question with the aid of improved methods for recording the production of gastric juice have shown moreover that histamine, when injected intravenously, produces copious secretion in the stomach. Lim found that blood from the fed animal has no effect upon gastric secretion when transfused into the circulation of another; and on these and other grounds denies that there is a gastric excitant in the blood after feeding. On the whole the balance of evidence now available does not warrant the conclusion that internal secretion plays any part in gastric secretion.

REFERENCES

BAYLISS and STARLING. *Journ. Physiol.* 28, 29, 1902–3.
DALE and LAIDLAW. *Journ. Physiol.* 44, 1912.
EDKINS. *Journ. Physiol.* 39, 1906.
LIM and co-workers. *Quart. Journ. Exp. Physiol.* 13–15, 1923–5.
PARSONS. *Am. Journ. Physiol.* 71, 1925.
STEPP. *Journ. Physiol.* 43, 1912.
MELLANBY. *Proc. Phys. Soc., Journ. Physiol.* 61, 1926.

§ II

Turning from the rôle of endocrine factors in the regulation of the digestive secretions, we shall now discuss a hypothesis recently put forward by Dixon and Marshall (1924) to account for the phenomena attending the termination of pseudopregnancy in the mammal. Many speculations have centred round the question "as to why" in the words of Michael Foster "the uterus after remaining for months subject to futile contractions is suddenly thrown into powerful and efficient action, and, within a few hours or even less, gets rid of a burden which it has borne with such tolerance for so long a time." The hypothesis of Dixon and Marshall may be summed up as follows: the occurrence of a certain stage in the cyclic activity of the ovary is a factor in parturition; the secretion of the ovary in the absence of fully formed corpora lutea has a specific effect in promoting pituitary secretion, which in its turn has a specific excitatory action on the uterus.

The explanation put forward by Dixon and Marshall, one which its authors do not claim to account for all the known facts, is clearly a proposition that claims the utmost consideration as offering a straightforward and reasonable solution of this obscure problem. In order to separate contributory evidence founded upon well established observations from propositions concerning which there is room for a certain element of uncertainty, it will be convenient to discuss the merits of this hypothesis under three headings: (1) evidence for the existence of an active substance in the mammalian pituitary which acts as a specific stimulant upon uterine muscle; (2) evidence that such a substance is actually discharged into the body fluids; (3) evidence that it is discharged into the body through the agency of ovarian secretion.

(i) Let us first consider the evidence for the existence in the pituitary of a substance which acts as a specific stimulant to uterine muscle. The reaction of uterine muscle to pituitary extract was first studied by Dale (1909), who showed that in contrast to the response to adrenaline the uterine muscle of the mammal invariably reacted to pituitary extract by tonic contraction even when stimulation of the hypogastric nerves produces pure inhibition of tone and rhythm; and that ergotoxine which abolishes the excitatory effect of adrenaline may be given in any quantity without affecting the contraction of uterine muscle in the presence of pituitary extract. The substance responsible for the effect is present in the posterior lobe of the mammalian gland, more particularly in the pars nervosa. It is extremely soluble in water, can be extracted in cold water or saline and withstands boiling. Dale showed that it is readily hydrolysed with trypsin and oxidised by hydrogen peroxide. It is undestroyed by peptic digestion. Of its chemical properties more details will be given in another chapter.

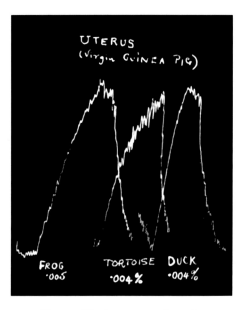

Fig. 21. Virgin uterus, guinea-pig.

As attempts to destroy or remove the posterior lobe of the pituitary in the mammal have not so far thrown any very definite light upon its functional significance, further evidence on this question must be sought in data relating to the possible presence of a substance in the body fluids identical with the active constituent of pituitary extracts. This necessitates accurate methods for quantitative estimation of the activity of the latter. Fortunately for our present purpose no method of biological assay is more suitable to accurate requirements than the uterine test first described by Dale and Laidlaw (1912) and later elaborated somewhat by Burn and Dale (1922). In competent and experienced hands it should, judging from the writer's own experience, give results correct to 5 per cent., though needless to say a thorough knowledge of the technique is necessary if satisfactory results are desired. As the quantitative aspect of work in this field is too often overlooked by endocrinologists, it may be as well to give the full details of the method as described by Burn and Dale.

The horn of a suitable uterus having been carefully dissected and suspended in the bath of warm Ringer's solution must be left to itself till relaxation is approximately complete. This will usually take from 20 to 30 minutes. Any necessary adjustment of the weight of the lever is then made. A first dose of the standard preparation is then given and the effect recorded. As a first dose we ordinarily use 0·5 c.cm. of a fifty-fold dilution of the standard fresh extract...or 0·01 per cent. extract [dry weight]. We start then with a dose of 0·5 c.cm. of 0·01 per cent. extract added to the bath containing 125 c.cm., giving a dilution of one part of original dry weight in 2½ millions. The effect varies much with different samples of muscle. With an initially sensitive preparation the result will be a contraction to a maximum maintained for a few minutes. In such a case the contents of the bath are changed for fresh Ringer's solution and time allowed for complete relaxation. This process after such a first dose is frequently less rapid than at a later stage. After 15–20 minutes, however, the muscle should be relaxed and ready for a second dose which in such a case is smaller than the first. After a few trials a dose is found—usually 0·3 to 0·4 c.cm. of the indicated solution—to which the muscle responds regularly, with identical contractions which are less, but not widely less, than the maximum of which it is capable. In the case of a less sensitive preparation, especially with the delicate horn from the uterus of a very young guinea-pig, the first effect of 0·5 c.cm. of the dilution may be very small. In such a case it is well, after change to pure Ringer's solution and relaxation, to repeat the same dose. It will usually produce a much larger response. If after three or four such doses the response is still widely submaximal, the dose must be increased....The aim in any case is to establish for the particular preparation, the dose which is about 60–70 per cent. of that needed to produce contraction to the maximum. When this dose has been fixed, it is essential to make sure, not only that the response to it is regular, but that an increase or decrease of the dose by 20 per cent. produces a clear increase or decrease in the response. By this stage of the test it will have become necessary to fix the time interval between the doses. This should be so chosen as to allow complete relaxation and persistence of the relaxed condition for about 2 minutes before the next dose is given. The suitable interval will vary with different preparations from 8 to 12 minutes, with an average of about 10 minutes. When once chosen it must not be arbitrarily or irregularly varied, though the change in the rapidity of the reaction and relaxation may make desirable a definite shortening of the interval at a later stage in the test. It will be seen that these preliminary tests to fix the suitable submaximal dose, and to demonstrate the possibility of accurate differentiation at this level of dosage, will necessarily occupy an hour or two at the rate of 5 to 6 doses per hour. Similarly when the standard dose is fixed the determination of the exactly equivalent dose of the preparation

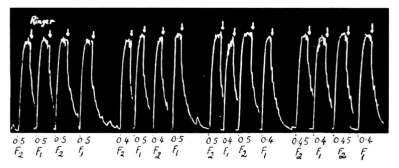

Fig. 22. A comparison of two extracts, F_1 and F_2. In each case for testing, a 50-fold dilution of a 2·5 per cent. extract of the fresh material is used, and the figures under the tracing represent the amount of each extract in c.c. added to the bath of 125 c.c. The tracing shows our regular procedure in comparing two preparations. After finding doses producing equivalent contractions of the uterine horn, as in the first four seen above, we then demonstrate that by slightly varying the doses contractions of correspondingly different heights are produced.

to be tested cannot be made quickly. In fixing this equivalence we alternate doses of the unknown with those of the standard extract, keeping the same constancy of time interval. A few sighting shots suffice to indicate the range within which the appropriate dose will be found, and the discrepancy is gradually narrowed by changing the dose either of the standard or of the unknown extract, until doses are found in which the two given in an alternate series of 3 or 4 doses produce curves of exactly the same height on the record. It is then desirable to make sure that reduction or increase of the dose of the unknown produces curves definitely lower or higher than that produced by the fixed dose of the standard (Fig. 22)."

It may be added that it is essential that the guinea-pig should be non-oestrous as well as virgin, that Dale's saline medium should be used, and the temperature carefully controlled.

The most active preparation of the uterine substance which has been obtained hitherto has been prepared in Abel's laboratory. The posterior lobes of bovine glands are dessicated and, after removal of fat, ground with solid mercuric chloride and 0·2 per cent. HCl. The mixture is shaken for several hours and the insoluble "cake" is washed twice by grinding with saturated mercuric chloride solution, and filtering with suction. A suspension of this "cake" in water is neutralised and treated with H_2S, saturated sodium chloride being added, to flocculate the insoluble mercuric sulphide. The suspension is filtered and aerated to remove excess of hydrogen sulphide. After decomposition of the mercuric compound a further precipitation by phosphotungstic acid is (equal volumes) effected. This precipitate is reprecipitated by re-solution in sodium carbonate solution and addition of hydrochloric acid. The reprecipitated phosphotungstate is finally taken up in dilute sodium carbonate and treated, ice cold, with hot saturated $Ba(OH)_2$ drop by drop till precipitation of the phosphotungstate is complete. Excess of barium is removed by H_2SO_4 in the usual way. The filtered solution is then treated with tannic acid and saturated with NaCl. The tannate precipitated is ground with 95 per cent. alcohol and a small quantity of tartaric acid. The alcoholic tartaric acid extracts are reprecipitated several times with large excess of anhydrous ether. The tartrate is dissolved in water and treated ice cold with saturated picrolonic acid in alcohol. An inactive precipitate is thus separated. This process is repeated. The clear solution is evaporated, redissolved in 93 per cent. alcohol, precipitated with much ether, and dried. This tartrate can be further purified by organic solvents. Dudley has also recommended the following procedure for obtaining preparations of a high degree of potency and comparatively free from proteins. The aqueous extract is acidified and filtered. Collodial ferric hydroxide is added in just sufficient quantity to precipitate proteins, leaving the supernatant fluid colourless. This suspension is now filtered, the active material remaining in solution. By continuous extraction with butyl alcohol from its aqueous solution the oxytocic substance can be completely separated. The butylic extract can be subsequently evaporated and concentrated aqueous solutions obtained therefrom. By repeated extraction from aqueous solution it would seem to be possible to obtain an extract which has only the oxytocic action of pituitary extracts; but the discussion of this point will be taken up at a later stage.

The most arresting feature of this response is its extraordinary

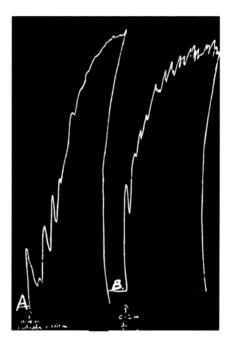

Fig. 23. Pituitary tartrate 1250 times stronger than histamine acid phosphate. Contraction of the virgin guinea-pig's uterus, suspended in 30 c.c. Tyrode solution, produced by a 1 : 250,000,000 solution of a pituitary tartrate and a 1 : 100,000 solution of histamine acid phosphate. A. 0·4 c.c. pituitary tartrate (0·0000016 mg. or 1 : 18,750,000,000). B. 0·2 c.c. histamine acid phosphate (0·002 mgm. or 1 : 15,000,000). With a more sensitive uterus, with which only 0·0005 mg. histamine acid phosphate was required to produce a vigorous contraction, probably only 0·0000004 mg., or 1 : 75,000,000,000, of this powerful tartrate would have produced an equally strong contraction.

sensitivity. In his most recent paper Abel (1924) describes a positive reaction in the guinea-pig's uterus with his "tartrate" preparation of the active compound at a dilution of 1 in about 20,000,000,000. This preparation is about 12,050 times as active as histamine acid phosphate. This means that 1 gramme of posterior lobe substance cannot contain more that 0·0002 gm. of the active substance. When it is remembered that this tartrate is a mixture of substances, it is clear that the uterine stimulant or, as it is sometimes called, *oxytocic* constituent of pituitary extract far surpasses in activity any known drug. The oxytocic substance is present according to Hogben and de Beer (1925) in the pituitaries of the Bird (duck), Reptile (tortoise), Amphibia (frog), Teleost (cod), and Elasmobranch (skate). The amount however present in the skate's pituitary is in terms of equivalent dry weight of gland substance very much smaller than in the pituitary of the cod. (Cf. Fig. 31.)

In respect of the extreme delicacy of the reaction of the uterus to pituitary extract the oxytocic test is unquestionably specific. For whether the effect of the pure oxytocic substance will prove, when isolated, to be restricted in its action to uterine muscle or not, there is no doubt that the sensitivity of the uterus to its influence is of an entirely different order of magnitude from that of other plain muscle. Dale in his original researches did not find that pituitary extract had any significant effect on the intestines and urinary bladder, though he concluded that it acted on the plain muscle of the arteries, a view which will be discussed in another place. However, in spite of Dale's observations, most other investigators with the notable exception of Guggenheim have assumed that the oxytocic substance is not specific for uterine muscle. Hogben and Schlapp (1924) who have emphasised the necessity of purifying pituitary preparations from histamine-like and choline-like substances present in all commercial extracts have expressed doubt as to whether there is any satisfactory proof that pituitary extract exerts a generalised action on plain muscle other than that of the uterus: and recently the matter has been very thoroughly re-investigated by Macdonald (1925) who has paid special attention to this point.

Apart from the presence of various non-specific plain muscle stimulants, it is important to bear in mind that saline commercial extracts of pituitary are of acid reaction, so that it is most important to keep a check on the pH of the solutions employed. In view of the difference in method and technique, which different investigators have employed, the divergent observations recorded in regard to extracts of the pituitary on intestinal muscle need occasion little surprise. Of those who have used laboratory products Atwell and Marinus (1918) and Shamoff (1916) claim that pituitary extract induces inhibition even at a dilution of 0·0002 per cent. Young (1915) and Guggenheim (1914) regard excitation as the characteristic reaction. But Young who found the effect to be inconstant on the rabbit found it necessary to use concentrations

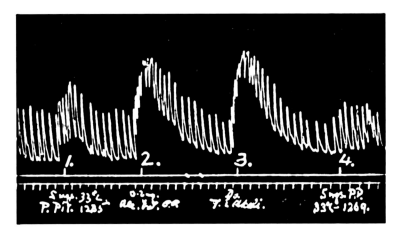

Fig. 24. Showing the response to 5 mg. gland and to 0·2 mg. of the residue obtained from a Soxhlet extract with absolute alcohol (six hours) before and after alkali digestion. (1) 5 mg. desiccated posterior lobe hypophysis. (2) 0·2 mg. residue on alcoholic extraction. (3) The same, boiled for thirty minutes with decinormal NaOH and neutralised before application. (4) 5 mg. desiccated posterior lobe hypophysis. (2) and (3) are of the same order of magnitude, hence the intestinal stimulant not only differs from the pressor and oxytocic principles in being alcohol-soluble, but is not destroyed by chemical processes which these principles cannot withstand (Macdonald).

of 0·05 to 0·1 per cent. to produce this effect on the cat's intestine; and Guggenheim showed that the exciting substance, *unlike the oxytocic* autacoid, is not destroyed by boiling with sodium hydroxide. Degener (1922) has even combined these divergent views in the conclusion that the response to extracts of the pars nervosa is relaxation while to those of the glandular portion it is contraction, both lobes giving the latter when the subject is fed on oatmeal and milk. Reference to other accounts based on the use of commercial preparations will be found in Macdonald's paper. Macdonald finds that with proper attention to reaction etc. most pituitary extracts do stimulate intestinal muscle, but that relatively high concentration is required, and the amount of intestinal activity present is not greater than that obtained from other tissues (brain, liver, spleen, thyroid etc.). As with the histamine-like substance which produces the depressor effect on the

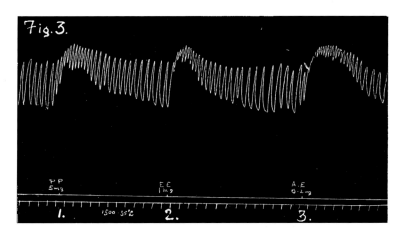

Fig. 25. Showing the response of a slip of ileum to saline extracts of gland substance, and the residues obtained on evaporating ether and alcohol extracts of the same. (1) P.P., 5 mg. desiccated posterior lobe hypophysis, previously extracted with ether and absolute alcohol. (2) E.E., 1 mg. of the dried residue of an ether extraction of gland in a Soxhlet apparatus (six hours). (3) A.E., 0·2 mg. of the dried residue from an extraction (Soxhlet, six hours) with absolute alcohol. It is to be noted that (2) E.E. is practically free from both pressor and oxytocic principles, while (3) A.E. contains only the "depressor" principle, if tested on an etherised cat (Macdonald).

circulation, the intestinal stimulant is extracted by ethyl alcohol which leaves the oxytocic substance behind. He confirms Guggenheim's statement that the oxytocic action of pituitary extract can be destroyed by sodium hydroxide leaving the intestinal stimulant unaffected. He draws attention to the great sensitivity of the mammalian gut to such vasodilator substances as histamine and choline. There is no specific action of pituitary extract on cardiac muscle. To sum up the main points of Macdonald's argument

pituitary extracts vary greatly in their content of intestinal stimulant and independently of their oxytocic activity; the intestinal
stimulant differs from the other known pituitary autacoids in being
alcohol-soluble and alkali-stable; extracts of laboratory preparations
do not act in concentrations to which physiological importance
need be attached; and, in any case, pituitary extract is not materi-
ally richer in intestinal stimulant than several other tissues. Since
pituitary extracts have been shown by Waddell to have no action
on the frog's oesophagus, and by Hogben and Hobson (1924) to
produce no response on the plain muscle of several invertebrates
(heart of *Pecten*, crop of *Aphrodite* and *Aplysia*), the conception of
the oxytocic substance as a general excitant of plain muscle requires
revision.

(ii) Having dealt briefly with the existence of an uterine stimulant
in the pituitary gland, we may now turn to ask whether this substance is discharged in the body fluids. For critical investigation
of this issue a great service has been done by A. J. Clark and his
collaborators who have emphasised the need for careful quantitative
study of the plain muscle excitants in blood, ymph, cerebro-spinal
fluid and tissue extracts generally. An extensive literature exists
on the presence of plain muscle excitants in the blood stream.
Fresh uncoagulated blood has a feeble vasoconstrictor action, a
potent effect upon the surviving uterus (Stewart and Zucker),
and variable effects (often combined inhibition and stimulation)
on the mammalian gut (Clark and Gross). Coagulation greatly
intensifies these activities (O'Connor, Richardson and Clark, Jane-
way, *et al.*). The substance (or substances) which produces this
effect is soluble in water and alcohol and insoluble in ether; and
resembles in several respects the action of the alcohol-soluble
substance obtained from peptone (Clark). Such facts are of the
utmost significance in dealing with the evidence for the question
which we are about to discuss.

In 1908 Herring put forward a hypothesis based on a study of
the histology of the pituitary throughout the vertebrate series.
According to this the active substances of the posterior lobe are
formed by disintegration of the cells of the pars intermedia to form
hyaline bodies, which in some way or other migrate into the pars
nervosa, whence the autacoids diffuse into the cerebro-spinal fluid.

Whether histological data such as those which Herring presents can have a very close bearing on conclusions of this nature is a question about which physiologists would be not unanimous. In any case a final decision must await the issue of quantitative experiment. Its chief interest for our present purpose is that in seeking for evidence for the discharge of the pituitary autacoids attention has been almost exclusively focused on the cerebrospinal fluid.

It must be admitted that the evidence on this question is conflicting. The first attempts to verify Herring's hypothesis were made by Cushing and Goetch (1910), who confined their attention to the pressor reaction. The results of Carlson and Martin (1911) and of Jacobson (1919) are entirely contradictory: they find no specific pressor effect. Cow (1915) and Dixon (1922) re-affirm the presence of a vasoconstrictor substance and a plain muscle excitant which is like the oxytocic substance destroyed by trypsin. But both these authors find that the action of the cerebro-spinal fluid on the uterus and the intestine is comparable, which would rather suggest that the oxytocic action of the cerebro-spinal fluid has nothing to do with the presence of the pituitary autacoid. Guggenheim's test was not employed. None of these authors have compared quantitatively the reaction to blood and cerebrospinal fluid. This has recently been carried out in Clark's laboratory. De Beer, Dreyer, and Clark (1925) have compared the vasoconstrictor, melanophore and oxytocic action of the blood and cerebro-spinal fluid of dogs using Dixon's technique. The melanophore stimulant is not present in cerebro-spinal fluid. The oxytocic activity of the normal cerebro-spinal fluid is of the order 0·0005 per cent. fresh posterior lobe substance, as compared with an average oxytocic activity of fresh dog's serum of the order 0·003 per cent. These figures are in substantial agreement with those of Dehane (1918). The results of recent work by Trendelenberg (1924), who apparently believes that this reaction is due to traces of the pituitary autacoid in the cerebro-spinal fluid, cannot however be taken as corroborative of Dixon's observations, for he finds that 1 c.c. of the cerebro-spinal fluid of cats is equivalent in oxytocic activity to 1/35,000 mg. fresh posterior lobe tissue as against Dixon's evaluation of 0·2 to 2 mg. for 1 c.c. of the cerebro-spinal fluid, of normal dogs. While the data derived from quantitative study of normal cerebro-spinal fluid cannot be said to furnish conclusive evidence in favour of the belief that the uterine autacoid is discharged from the pars nervosa into the cerebro-spinal fluid, there are certain observations recorded by Dixon (1923) and others by Trendelenberg (1924) which must be carefully reinvestigated before this hypothesis is rejected. Dixon states that injection of pituitary extract (commercial products) into a normal animal is followed by an increase in the oxytocic activity of the cerebro-spinal fluid; and he finds that this increase does not occur after destruction of the pituitary. From this he concludes that pituitary extract stimulates the pituitary to secretory activity in much the same way as bile salts excite the liver. However Dixon admits that the cerebro-spinal fluid still gives the oxytocic test after hypophysectomy; and Clark and his collaborators could not substantiate the claim that injection of pituitary extract increases the oxytocic activity of the cerebro-spinal fluid. Trendelenberg, on the other hand, finds that it decreases appreciably 24 hours after transection of the stalk of the gland in the Mammal.

(iii) For a discussion of the remaining question—the relation of the ovary to pituitary secretion,—it is unfortunate that there is so little on which to base an affirmative answer to that which has just been proposed. For the whole argument turns upon whether the oxytocic activity of the cerebro-spinal fluid is due to the discharge of the specific uterine stimulant present in the pituitary gland.

Dixon (1922) claimed that of all tissue extracts which he employed those of the ovary—and the posterior lobe itself—alone increase the oxytocic activity of the cerebro-spinal fluid. Extracts of corpora lutea and liquor folliculi have not this action. Subsequently Dixon and Marshall have compared the oxytocic activity of the cerebro-spinal fluid of various domestic Mammals and are able to correlate its intensity with stages in the oestrous cycle of the female, and the course of normal pregnancy and parturition. The conclusions of these authors are best stated in their own words:

Our experiments consistently support the view that in the presence of fully formed and presumably functional corpora lutea the normal ovarian secretion is largely or entirely in abeyance, and this is the condition for a short part of the time between the heat periods, but more particularly during pregnancy. In other words the corpus luteum may be supposed so to dominate the ovarian metabolism at those periods that the ovarian secretion which at other times activates the pituitary is inhibited or else is neutralised by the secretion coming from the corpus luteum.

This fruitful hypothesis offers a reasonable explanation of the increasing irritability of the uterus in the later stages of pregnancy, but much more quantitative information is needed before it can be regarded as proven.

REFERENCES

ABEL. *John Hopkins Hosp. Bull.* **35**, 1924.

BURN and DALE. *Reports on Biological Standards, I, Pituitary Extracts. Medical Research Council Pub.* 1922.

DALE. *Biochem. Journ.* **4**, 1909.

DE BEER, BREYER and CLARK. *Arch. Intern. Pharm. Ther.* **30**, 1925.

DIXON. *Journ. Physiol.* **57**, 1923.

DIXON and MARSHALL. *Journ. Physiol.* **59**, 1924.

HOGBEN and DE BEER. *Quart. Journ. Exp. Physiol.* **15**, 1925.

HOGBEN and HOBSON. *Brit. Journ. Exp. Biol.* **1**, 1924.

MACDONALD. *Quart. Journ. Exp. Physiol.* **15**, 1925.

Chapter V

THE RELATION OF INTERNAL SECRETION TO VASOMOTOR REGULATION

§ I

The characteristic effects of intravenous injections of pituitary extracts upon the circulation were first recorded by Oliver and Schafer (1894) simultaneously with their observations on the action of suprarenal extracts. Since that time the possibility that endocrine factors might play a part in the control of the circulatory system itself has been in the minds of numerous investigators. The co-operation of the chromaphil tissues in vasomotor regulation may now be dismissed, at least it is improbable that the concentration of adrenaline in the blood ever rises to a level at which it is capable of exerting any detectable effects upon the blood pressure. The possibility that pituitary activity is concerned in the normal reactions of the peripheral vessels has on the other hand assumed a new interest through recent work of Krogh which illustrates a rather different method of attacking the problems of internal secretion from those which have been discussed hitherto. For the benefit of the zoological reader it may be advisable to pause at this point and define the problem of vasomotor regulation.

The primary function of the circulatory system of animals is to respond to what Barcroft terms the call of the tissues for oxygen: other things being equal, the rate at which a tissue can take up oxygen depends upon the amount of blood which flows through it in unit time. The flow of liquid through a tube depends upon the force propelling it—the intermittent action of the heart in this instance, the viscosity of the fluid (an important but little known aspect in the case of the circulation), the length of the tube which can here be regarded as constant, and the sectional area of the tube which, owing to the presence of contractile elements in the walls of the blood vessels, is a variable quantity. Relaxation and contraction of the contractile elements investing the walls of the vessels supplying an organ may be brought about by extrinsic agencies such as nervous stimulation, autacoids and drugs, or by the local action of metabolites produced during the activity as illustrated so well by the work of Barcroft on salivary secretion. Stimulation of the central ends of most sensory nerves (other than the vagus depressor) produces reflex rise of blood pressure in Vertebrates via the vasoconstrictor paths which diverge from a centre in the bulb. Constriction of arterioles is brought about by peripheral stimulation of sympathetic fibres whose cell stations are situated in the chain ganglia. Stimulation of afferent nerves produces a localised dilation which is not reflex in origin, but appears to depend on the fact that certain sensory fibres have motor terminations in connexion with the contractile elements of the finer vessels. The condition of the vessels has been studied chiefly by three methods: (i) manometric measurement of blood pressure, (ii) volume changes of individual organs (plethysmograph or oncometer), (iii) rate of flow from open vessels in artificial perfusion experiments. The recognition through the work of Krogh that the capillaries are active agents in determining the peripheral resistance to the blood flow implies that none of these methods is wholly satisfactory; and our views on the nervous regulation of the blood flow may undergo some revision in details in the near future.

Until quite recently the capillaries had been regarded by physiologists as simply the organs concerned with the interchange of substances—in particular, of course, oxygen and carbon dioxide—between the blood and the tissues. They were not regarded as taking any active part in determining the rate at which the blood flows through them, i.e. the rate at which oxygen can be delivered to the tissues. It was assumed that the capillaries reacted in virtue of their elasticity alone to the state of contraction or dilation of the arterioles. Among the observations which have led to a revision of this view the work of Dale and Richards (1918), who sought to analyse the action of three depressor drugs, may be mentioned. Since the publication of Krogh's direct observations on the capillary system of the frog, the conclusion that capillaries as well as arterioles are capable of independent contractility has gained an increasing number of adherents. Any difficulty arising from the absence of muscular fibres in the walls of the capillaries now seems to have been removed by the observations of Vintrup (1922) who has demonstrated the contractile nature of certain anastomosing cells originally described by Rouget (1873) whose work was for nearly half a century disregarded[1].

For a summary of the evidence for independent contractility of the capillaries the reader is referred to Krogh's monograph. The issue which concerns us here is Krogh's hypothesis that the pituitary secretes a hormone which is a normal agent in the maintenance of a certain measure of capillary tone. Krogh's investigations into this question originated from the experience that though the capillaries in the web of the frog can be obtained in a state of normal tone when perfused through the femoral artery with blood for periods of forty minutes to two hours, perfusion with gum Ringer containing dialysed blood corpuscles leads rapidly to dilatation of the capillaries and stasis with recovery when normal blood is substituted. With a Ringer dialysate of blood, this phenomenon did not occur. It followed that there is present in normal blood a dialysable constituent on whose presence the preservation of capillary tone depends. Active dialysate from mammalian blood can be preserved on ice without losing its activity: it is not destroyed by boiling; it is not removed by ethyl alcohol or ether. There have been so many seemingly aimless investigations in the field of endocrinology that it is not without interest at this point to quote Krogh's comment on the search for the nature of the active substance. "I had sometimes thought," he writes, "of testing a number of extracts from glands known to possess internal secretion, in the hope of finding X in one of them; but I have always had a strong dislike for experimentation at random, and no tests of the kind were carried out."

The clue was suggested by Pohle's (1920) observations on the development of cutaneous oedema after extirpation of the pituitary in frogs. This oedema is a very characteristic post-operative phenomenon in adult frogs, though not in hypophysectomised larvae. Oedema is dependent on a state of capillary dilatation; and the obvious inference was put to the test of direct observation on the capillaries of the web by Krogh and Rehberg (1922). The results of removing the gland fully confirmed their expectations. A few hours after removal of the whole pituitary the capillaries of

[1] This view has recently been criticised by Floray and Carleton, *Proc. Roy. Soc.* B 100, 1926.

the web widen, and attain maximal dilatation after about twenty-four hours. As the experiments of Krogh and Rehberg have not yet been independently confirmed, the present writer may take the opportunity of stating that in experiments on the control of colour response which were carried out about the same time as those of Krogh and Rehberg this condition was observed in over a hundred and fifty hypophysectomised frogs. On microscopic examination the web presents an appearance never witnessed in the normal animal owing to the exaggerated engorgement of the capillary bed. Simultaneously the palm of the fore limbs assumes (in *Rana temporaria*) a noticeably pink hue for the same reason. According to Krogh, frogs from which the hypophysis is removed do not survive the operation for more than a few weeks, and the capillaries regain their contractility after a week or a fortnight. Using a different technique for the removal of the gland, the present writer has been able to keep hypophysectomised frogs alive for several months, during the whole of which time the characteristic capillary disturbances persisted. The removal of the "whole" gland in the frog (i.e. pars anterior, intermedia, and nervosa) does not involve the extirpation of the pars tuberalis which is represented by two glandular elements lying quite detached on the sides of the tubes. As the capillary disturbance is quite ephemeral when the pars anterior alone is removed, it is clear that the pars intermedia or nervosa alone are involved.

Injection of a drop of pituitary extract into the web of the frog's foot brings about contraction both of the capillaries and of the smaller arterioles. It has long been known that perfusion with a solution containing pituitary extract in quite minute quantities results in constriction of the peripheral circulation of the frog. By direct observation of the web and with cinematographic records Krogh has shown that visible effects on the capillaries alone are produced when the vessels are perfused with a Ringer containing a concentration of one in a million commercial pituitary preparation. Assuming the latter to be a 10 per cent. extract, this represents a dilution of 1 : 10,000,000 fresh glandular substance. The extreme sensitivity of the capillaries for "pituitrine" removes any considerable difficulty in bringing the results of hypophysectomy into harmonious relationship with the effects of perfusion of pituitary extract on the circulatory system.

Krogh's researches present a strong case for the hypothesis that in the frog at least the maintenance of capillary tone depends to some extent upon endocrine agencies. The question arises whether this conclusion is applicable to the Mammal in which the vaso-constrictor action of pituitary extracts was first demonstrated. It may as well be said that at present no definite answer can be given to this question. On the other hand, further possibilities are opened up by Krogh's work in the light of which the analysis of the vaso-motor properties of pituitary extract becomes more than ever a field for useful work. And in judging the applicability of Krogh's hypothesis to the Mammal, it is clearly of interest to know how these properties manifest themselves in other groups of vertebrates.

REFERENCES

KROGH. *The Anatomy and Physiology of the Capillaries.* Yale. 1922.

§ II

Judging by its effect upon the blood pressure pituitary extract does not seem to have much influence upon the peripheral circulation of the Reptile (tortoise). If it has any action it is predominantly vasodilator (Hogben and Schlapp). But in the frog which responds to very minute quantities of pituitary extract when the vessels are perfused, relatively large quantities are required to produce an appreciable rise of the blood pressure in the pulmo-cutaneous artery.

Concerning the vasomotor action of pituitary extract in the case of the Bird there is however no doubt. Pituitary extract in any quantities which produce a manifest effect on the carotid blood pressure evokes a predominantly depressor response. This was first shown by Paton and Watson (1912) in the decapitated duck. Their experiments showed that both initial and subsequent injection of commercial pituitary extract resulted in a steep fall in blood pressure chiefly, as plethysmographic records demonstrated, through dilatation in the splanchnic area. These observations were however based upon commercial preparations that are known to contain substances which, as will be seen below, exert a depressor

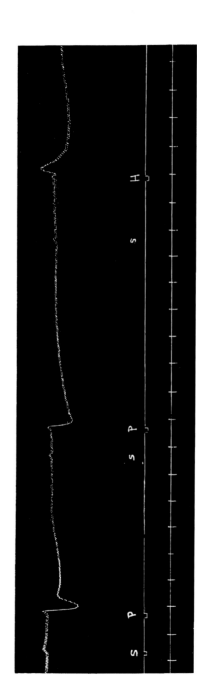

Fig. 26. Duck (urethane-ether), carotid blood-pressure. S, 1 c.c. saline; P, 1 mg. sample B ox posterior lobe; H, 1 mg. ergamine phosphate.

action on the mammalian circulation. Hogben and Schlapp (1924) have however repeated Paton and Watson's experiments, using laboratory products from which these depressor substances had been removed. From these observations it appears that the substance responsible for the avine depressor response is clearly not histamine nor choline nor histamine-like substances present in commercial preparations and in other tissue extracts. It has in fact nothing to do with the secondary fall to be discussed later in relation to the Mammal.

More recently the question has been reinvestigated by Hogben (1925) who finds that the avine depressor substance is a specific property of fresh pituitary extract: the depressor response is not evoked by fresh extracts of other organs (spleen, liver, ovary, brain). The avine depressor component is confined to extracts of the posterior lobe of the bovine gland; and is present in the pars nervosa in relatively greater amount (per unit weight of gland substance) than in the pars intermedia. The fact that it is present in extracts of the anuran hypophysis indicates that it is not formed in the pars tuberalis. It is also present in the pituitaries of the Birds (duck), Reptile (tortoise), Fish (cod and skate). It seems that extracts of the cod's pituitary are more potent than those of the skate's; but the precise ratio of activity was not determined. It does not pass readily through a collodion membrane: like the oxytocic, mammalian pressor and melanophore substances it is readily soluble in water, insoluble in ethyl alcohol, withstands boiling, and is readily inactivated by trypsin and alkali. Whether it is identical with the substance which produces a rise in blood pressure in the Mammal is uncertain. In the duck a pure depressor response is obtained even after the circulation of the splanchnic area is excluded. There is, as with the mammalian pressor response, an acquired tolerance to successive doses. In the fowl Hogben and Schlapp found some indications of a pressor response succeeding the steep initial fall. This was never seen in the duck.

REFERENCES

HOGBEN. *Quart. Journ. Exp. Physiol.* 15, 1925.
HOGBEN and SCHLAPP. *Ibid.* 14, 1924.
PATON and WATSON. *Journ. Physiol.* 44, 1912.

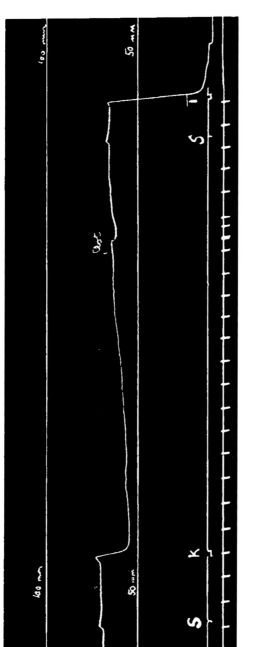

Fig. 27. Duck, urethane-ether. 29.xii.23. S, 1 c.c. s line; K, 10 mg. desiccated skate pituitary; T, 10 mg. desiccated cod pituitary. Time intervals, one minute.

§ III

Oliver and Schafer (1894-5) first described a prolonged rise in arterial blood pressure in the Mammal after intravenous injection of pituitary extract (Fig. 27 *a*). Howell (1898) showed that the autacoid was located in the posterior lobe and that later doses given at short intervals evoked responses which are less pronounced both in extent and duration. Schafer and Vincent (1898) described in the etherised dog a purely depressor effect after repeated injections. They attributed this reaction, and the slight preliminary fall which sometimes precedes the prolonged rise in response to the initial injection, to an alcohol-soluble substance which they showed was not choline. This view has been adopted by other workers including Miller, Lewis and Matthews (1911), Herring (1915), and Jacobson (1920). The depressor effect which is described in carnivora by these workers is not however seen in the rabbit. Nor have all authors accepted Schafer and Vincent's view. Dale and Laidlaw (1920) uphold the view that "the uninjured pressor substance can itself under proper experimental conditions (previous injection; age of animal; stage of anaesthesia) prepare the way for a decided fall of pressure in response to later injections." Bayliss (1922) again writes: "we have met with a reversal of this kind in the case of strychnine where the most satisfactory explanation was found to be that the process normally resulting in inhibition was reversed to an excitation...whether a similar explanation would hold for pituitrin; or whether the effect is more like the peripheral reversal effects mentioned below is uncertain."

The fact that a single "depressor" substance, histamine, evokes a purely depressor response in the etherised cat, and a predominantly pressor response in the etherised rabbit or spinal cat, suggests a basis for these misunderstandings. Hogben and Schlapp (1924) have shown that as with the response to histamine, the effects of successive injections of pituitary extract on the blood pressure of the spinal and etherised cat are quite different, if ordinary laboratory preparations or commercial products are used. If pituitary extract is subjected to *sufficiently prolonged* alcoholic extraction it never evokes a depressor response. When alcohol-extracted, "depressor-free" preparations are employed, the diminution of

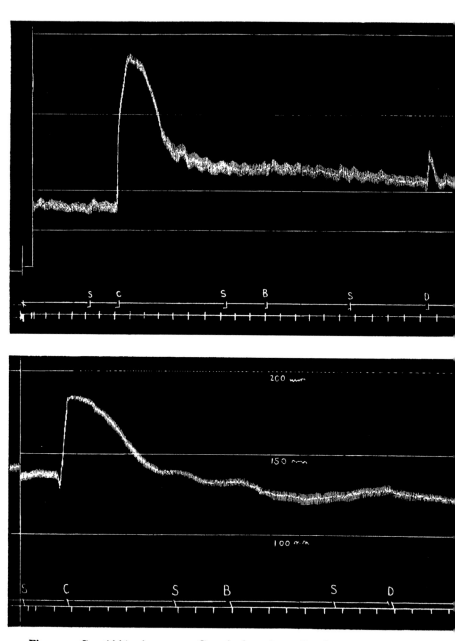

Fig. 27 *a*. Carotid blood pressure. Cat. A, decerebrate; B, etherised, both vagi cut. Inje
ox posterior lobe in 1 c.c. saline. Sample D extracted for forty-eight
All three samples gave the character

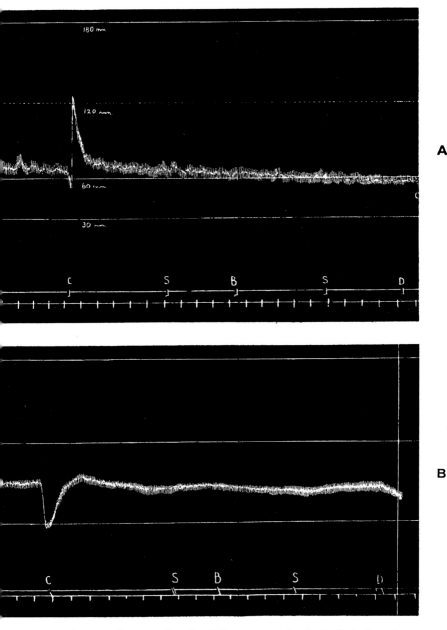

A

B

ut ten-minute intervals: S, 1 c.c. saline; B, C, D, 5 mg. samples B, C, D, desiccated
hyl alcohol, sample B for six hours, and sample C unextracted.
d rise on *initial* injection in this dosage.

response, though most obvious after the first, second, and third injections, is more or less continuous till complete immunity is attained, providing that the time interval is relatively short. After a lapse of time, varying with the size of the dose given, full response can be recovered when complete immunity has been produced by repeated injections; by administering the extract only at the end of the period adequate for complete recovery, a constant response can be obtained over periods of many hours. The following protocol is illustrative. It is to be noted that whereas the period of an hour and a half was required to restore the original response at the end of the experiment, a two hours' period was not sufficient after the succession of doses given within the first hour, thus suggesting that the amount actually circulating, and therefore possibly the rate of excretion, enters, as Dale (1909) first suggested, into the question as to how long a period is adequate for the recovery of maximum reactivity.

Spinal Cat. Operation at 10 a.m.

p.m.		mm.	mm.
1.20 systolic (carotid) pressure		—	52
2.0 4 mg. dried, depressor-free ox posterior lobe injected (jugular vein): pressure rises from 50 to 124 mm.		—	74
2.15 same dose repeated: pressure rises from		54–89	35
2.25	,, ,,	64–83	19
2.35	,, ,,	64–78	14
2.45	,, ,,	60–68	8
2.55	,, ,,	60–68	8
3.0	,, ,,	60–63	3
3.5	,, ,,	60–62	2

(The response to the last two doses being of the order of change produced by saline injection, complete tolerance was attained at 3 o'clock.)

5.5 same dose repeated: pressure rises from		60–126	66
5.35	,, ,,	63–86	23
7.36	,, ,,	54–128	74
8.26	,, ,,	54–110	56
10.0	,, ,,	54–127	73

For studying either the biochemistry or the functional significance, if any, of the pressor response in the Mammal it is clearly of the utmost importance to have a means of quantitative comparison of pressor activity. These observations indicate how this can be accomplished. They show in the first place that in order to make a valid comparison of pressor activity it is essential to remove the depressor substance or substances ordinarily present in extracts of pituitary. Second, they indicate that with sufficient intervals between injection, constant responses can be evoked by corresponding quantities over long periods of time. This is important, partly because in attempting to match samples on a diminishing response it is altogether impossible to get results of an order of accuracy comparable with those obtained in oxytocic standardisation; and also because in the spinal cat the "depressor" component may be the chief contributor to the pressor effects which are produced by successive injections. A method of standardising pressor activity with an order of precision not inferior to that obtained with the virgin uterus (10 per cent.) has been recently proposed by Hogben, Schlapp, and Macdonald (1925). The essential points in this method are as follows. For depressor-free extracts tolerance is a function of the dosage and time interval between successive injections. With appropriate time intervals a remarkable constancy of response of the same character and order of magnitude as the initial rise in blood pressure may be obtained with the spinal cat for periods up to 20 hours. With submaximal doses the period of recovery for response of the same order as that obtained on initial injection is rather less than an hour. Consistent discrimination tending to increase as the experiment proceeds may be obtained for ten or twelve hours between hourly doses differing by 10 per cent. The curve obtained by plotting increase of pressure against dose shows the steepest gradient near a point corresponding to half the dose requisite to produce maximal response; and the best discrimination is obtained by working with a standard in the neighbourhood of this value. It is therefore suggested that the match should be made against a standard for which a curve of reference is kept in use: the limit of accuracy can be defined in any instance by interposing a known dilution between two injections of the standard. In carrying out this method of standardisation rigidly the completion of an assay on one animal makes excessive demands upon the experimentalist. It is best therefore to make a rough comparison on one cat assisted by a standard curve relating dosage to rise in pressure. In the final

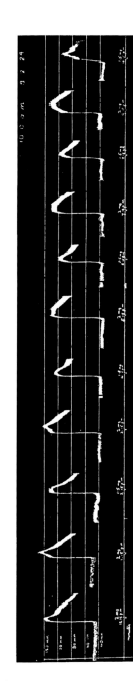

Fig. 28. Spinal cat. Showing the effect of eleven successive doses of pituitary extract at hourly intervals commencing at 12.15 p.m., and showing consistent discrimination between doses differing by 25 per cent. administered alternately after two injections of the standard. The operation was performed at 10 a.m. The time interval (bottom line) is one minute.

assay it is best to give seven doses, as follows: standard, 115 per cent. match, standard, match, standard, 85 per cent. match, standard. This enables one to state an outside limit of error in the assay: if the middle rise is equal to that on either side of it, the second greater than that on either side of it, and the sixth less than that on either side of it, the match is correct within 15 per cent. and the ratio of activity of match to standard is the inverse ratio of their dilutions for equal quantities injected. An alternative

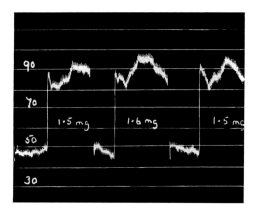

Fig. 29. Showing 7 per cent. discrimination after thirteen previous injections at hourly intervals. Sixteen hours after operation.

method of assay has since been proposed by Heymans (1925) who perfuses the vessels of the rabbit's head. It is doubtful whether great accuracy can be obtained by this procedure, though it has an advantage in that it is not so protracted.

Having indicated some of the difficulties which beset a quantitative treatment of the mammalian pressor response, it remains to enquire whether this specific property of pituitary extracts has a functional as well as pharmacological interest. At present there is no definite evidence from the attempts to extirpate the mammalian pituitary to throw any light on the problem. It does however clear the way for further enquiry, if we can settle whether the seat of action of the vasoconstrictor reaction in the Mammal is the same as in the frog, where there is good reason to attribute a physiological significance to the response. The view originally held by Schafer and by Dale was that the pressor response is an arteriomotor phenomenon; and this conclusion has been almost universally adopted. To the present writer it seems that there are quite inadequate grounds to warrant such an inference from the data available. And in the light of Krogh's researches it is at least desirable to re-examine the evidence for this view.

Among the reasons that have been advanced in support of the view that pituitary extract acts primarily on the muscular coat of

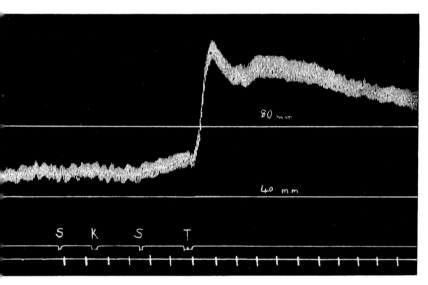

Fig. 30. Spinal cat, carotid b.p. S, 1 c.c. saline; K, 10 mg. desiccated skate pituitary; T, 10 mg. desiccated cod pituitary.

the arteries direct observations on surviving arteries deserve prior consideration. To mention only a few, De Bonis and Susanna (1909), Dale (1909), Campbell (1911), Cow (1913) have all recorded contraction of arterial rings in the presence of pituitary extract. At first sight this seems to be satisfactory enough. But since in none of these investigations pituitary extracts free of histamine-like substances or non-specific plain muscle excitants were employed, they cannot be said to carry any considerable weight. On the other hand, circumstantial support for the accepted interpretation has been provided by the tendency in the past to regard the oxytocic substance as an universal excitant of plain muscle and as identical with the pressor component. It has been seen in an earlier chapter that the former assumption is unwarranted. If it can be shown that the oxytocic and pressor substances are not identical, there remains no prima facie case for assuming, as is customary, that the pressor response is due primarily to arterial constriction rather than to capillary constriction as the work of Krogh suggests.

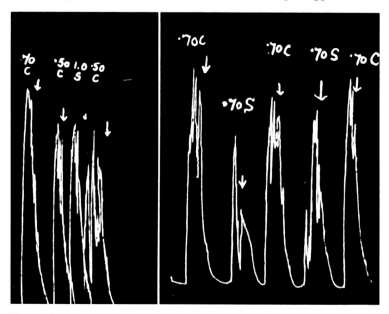

Fig. 31. Two comparisons of extracts of pituitary of skate and cod on uterus of virgin guinea-pig. Numbers indicate quantities in c.c. added to uterus bath. 1 c.c. C = 0·17 mg. of desiccated cod pituitary; 1 c.c. S = 2·2 mg. desiccated skate pituitary.

To do justice to this question leads to a consideration of some other aspects of the comparative physiology of the pituitary gland; for such indications as have been brought forward in regard to the separate identity of the pressor and oxytocic substances are drawn not only from the differential destruction and fractionisation of the two forms of activity by chemical means but also from their differential distribution among different species of animals and in different parts of the gland of the same animal. The latter may be dealt with first.

Herring (1908–15), who has carried out an extensive series of investigations on the comparative physiology and histology of the vertebrate gland, concludes from his investigations into the properties of pituitary extracts prepared from the glands of different Vertebrates: (*a*) that the pressor activity of posterior lobe preparations of the bovine gland is exclusively confined to extracts of the pars nervosa, whereas extracts of the pars intermedia display an oxytocic activity of an order of magnitude not greatly inferior to those of the nervous portion; (*b*) extracts of the pituitary of the Elasmobranch excite the uterus but exert no pressor action.

Phyletic Distribution of Pituitary Autacoids

	Elasmobranchii	Teleostei	Amphibia	Reptilia	Aves	Mammalia
Mammalian pressor	O (Herring) ? (Hogben and de Beer)	+ (Herring) +	+ (Herring) +	+ (Herring)	+ (Herring)	+
Avine depressor	+ (Hogben)	+ (Hogben)	+ (Hogben)	+ (Hogben)	+ (Hogben)	+
Uterine stimulant	+ (Herring) (Hogben and de Beer)	+	+ (Hogben and de Beer)	+ (Hogben and de Beer)	+ (Hogben and de Beer)	+
Melanophore stimulant	+ (Hogben and de Beer)	+ (Hogben and Winton)	+ (Hogben and Winton)	+ (Hogben and Winton)	+ (Hogben and Winton)	+
Diuretic action	O (Herring)	+ (Herring)	+ (Herring)	+ (Herring)	+ (Herring)	+
Galactogogue	+ (Herring)	+ (Herring)	+ (Herring)	+ (Herring)	+ (Mackenzie)	+

Herring alone has previously attempted quantitative comparison of the pressor and oxytocic activity of pars intermedia and pars nervosa. His figures with reference to the uterine response are in close correspondence with those obtained by Hogben and de Beer (1925). On the other hand, Herring did not find that the pars intermedia, when carefully separated, contains any pressor material. In denying the presence of the pressor substance in the pars intermedia, Herring has not received support from other investigators. Biedl originally located the source of the pressor substance exclusively in the pars intermedia. Both these diametrically opposed views are incompatible with the observations of Lewis, Miller and Matthews (1911), of Jacobson (1920), and of Atwell and

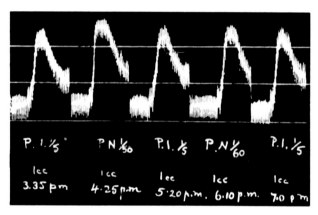

Fig. 32. Spinal cat, carotid b.p. The original extracts of pars intermedia (P.I.) and pars nervosa (P.N.) were equivalent to 10 mg. desiccated substance.

Marinus (1918), all of whom found that extracts of both the nervous and intermediate components yield extracts having both pressor and oxytocic properties. Atwell and Marinus add the additional information that extracts of the pars tuberalis are practically devoid of either. Herring lays great stress on excluding that part of the intermedia which borders on the pars nervosa: however, if this practice is too rigorously applied there remains little but the cone of Wulzen, which is histologically like the pars anterior. With the exception of Atwell and Marinus, the term pars intermedia has been used by those who have investigated the present issue to include indiscriminately three different tissues—pars intermedia in the comparative sense, cone of Wulzen, and pars tuberalis.

The matter has since been re-investigated by Hogben and de Beer (1925) with the use of the method of pressor standardisation outlined above so that it was possible to make comparative assay of pressor and oxytocic activity in identical samples of different parts of the same batch of glands with a comparable order of precision. While the experiments which they record do not support the belief that the pars intermedia is wholly devoid of pressor material, they are substantially consistent with the conclusion which Herring draws from his own researches, viz. that the distribution of pressor and oxytocic activity between the pars intermedia and nervosa is suggestive of the separate identity of the substances which evoke the uterine and pressor responses. On the other hand it may be questioned whether Herring's observations on the activity of extracts of the pituitary of the Fish are decisive. Herring observed that immersion of the uterus in a 0·25 per cent. extract of skate's pituitary evoked contraction, while injection of 3 c.c. of a 0·25 per cent. extract of skate's pituitary produced no rise in blood pressure in the Mammal. A dose of 1 c.c. of a 0·25 per cent. extract of ox pituitary (whole gland) should, it is true, give a fairly good pressor response; but it would not be of an order of magnitude very much greater than the threshold dose for a medium-sized cat. On the other hand, the final dilution in the uterus bath for oxytocic standardisation by Dale's method is of an order 0·0005 to 0·00005 per cent. Herring employed the rat's uterus, which he states to be less sensitive than that of the guinea-pig. But in the absence of more precise information, it is clear that further investigation is necessary before any conclusion as to the identity or non-identity of the pressor and oxytocic materials can be legitimately drawn from this experiment. Hogben and de Beer found that 50 mg. of desiccated skate's pituitary elicited no pressor response in a spinal cat which subsequently gave a pronounced and prolonged rise to 10 mg. of cod's pituitary. The skate's extract, if it possessed any pressor activity at all, was at best less than a tenth as potent as that of the cod. When however the uterine activity of the two samples was compared it was found that the cod's pituitary contained at least 20 and probably 30 times as much oxytocic material as that of the skate (Figs. 30 and 31). Hence no definite conclusions as to separate identity of the pressor and oxytocic substances can be derived from this source.

Researches on the chemistry of the active principles of posterior lobe extracts began with the discovery of the physiological action of the latter; but up to the present the results which have been obtained are somewhat disappointing. The main issue in relation to which it is possible to state any positive conclusions relates to the separate identities of the pressor and oxytocic autacoids and their relation to the melanophore stimulant.

The fact that all these forms of activities are destroyed by the same modes of treatment (tryptic digestion, acid hydrolysis, etc.) and precipitated by the same reagents (e.g. tannic and phospho-tungstic acids) at first sight suggests that they are manifestations of the presence of a single autacoid. Specially is this conclusion favoured by recent experiments of McClosky and Smith (1925) and of Schlapp (1925), who find respectively that the pressor and oxytocic substances diffuse at the same rate through collodion and that the pressor, oxytocic, and melanophore substances are destroyed at the same rate by acid hydrolysis. Thus Schlapp (1925) gives the following ratios for these activities in extracts submitted to treatment specified:

Acid Hydrolysis

(a) 1/10 Molar HCl.

	Unhydro-lysed	Boiled 20 min.	Boiled 40 min.
Oxytocic	100	50	19
Pressor	100	50	17

(b) 1/20 Molar acid.

	Unhydro-lysed	Boiled 40 min.	Boiled 80 min.
Oxytocic	100	·56	22
Pressor	100	63	21

(c) As (a) but deproteinised.

Oxytocic	100	41	12
Pressor	100	40	18

(d) As (c).

Melanophore	100	33	—
Pressor	100	30	—

On the grounds that the highly active compound whose preparation was described in chapter IV possesses all these attributes, Abel still maintains the conclusion that the pressor, oxytocic and melanophore stimulants are properties of a single active substance. It is the opinion of the present writer that the fact that Abel's tartrate gives the "inversion effect" (i.e. secondary fall) alone shows that it is not homogeneous. Three methods have been put forward as a means of separating fractions of pituitary extract in which the ratios of these several activities are significantly different. The first method was discovered by Dale and Dudley (1921), who found that the pressor substance can only be removed to the extent of 50 per cent. by continuous extraction from the watery solution with butyl alcohol, whereas the oxytocic principle can be completely removed by this procedure. Dudley (1923) has reaffirmed the conclusion that the solubility of the uterine and pressor components in butyl alcohol is significantly different; and has been independently corroborated by Fenn (1924). Any doubt arising from the unsatisfactory method of pressor assay adopted by these workers is removed by the observations of Schlapp (1925), who has employed the method of standardisation outlined earlier in this chapter. In three typical experiments the relative potencies of the alcoholic and watery layers and the original solution (control) are as under:

	Control	Alc. layer	Watery layer
Oxytocic	100	18	1·5
Pressor	100	20	14
Oxytocic	100	42	28
Pressor	100	11	63
Oxytocic	100	33	26
Pressor	100	6	71
Melanophore	100	3	71

A second method is due to Schlapp; and is based on absorption of the oxytocic substance to lead sulphide discovered in the first place by Guggenheim. When from an extract containing 0·01 per cent. of lead acetate, lead sulphide is precipitated by sulphuretted hydrogen, the extent to which the pressor and oxytocic substances are absorbed to the precipitate is significantly different. In Schlapp's experiments the following values were obtained:

(a)	Control	After absorption
Oxytocic	100	35
Melanophore	100	13
Pressor	100	10
(b)		
Oxytocic	100	31
Melanophore	100	8
Pressor	100	13

From the foregoing experiments by Schlapp there are no sufficient grounds for regarding the melanophore and pressor responses as manifestations of the activity of different components; but they indicate that the melanophore and pressor responses are due to the activity of a different substance from that which excites the uterus. The separation of the uterine and melanophore stimulants has also been recorded by Dreyer and Clark (1923), who find that the latter does not pass as readily as the former through collodion membranes.

In a more recent paper confirming the separate identity of the uterine and pressor principles Draper (*Am. Journ. Physiol.* **79**) calls attention to data given in Abel's own papers as corroborative of the foregoing.

To revert now to the question proposed at an earlier stage in this discussion it seems fairly certain that the oxytocic substance is neither a general stimulant of plain muscle nor identical with the substance which elicits the vasoconstrictor effects of intravenous injection of pituitary extract in the Mammal. Since direct evidence of the response of isolated arteries to pituitary extracts whose vasomotor properties are indubitably homogeneous is lacking, there is no longer any *prima facie* case for the view that the mammalian pressor response is predominantly arterial rather than capillary in origin. On the other hand, the undoubted depressor action of pituitary extract on the avine circulation makes it doubtful whether the conclusions derived from experiments on the lower Vertebrates can be extended to the Mammal. All we can say definitely is that Krogh's researches reveal new possibilities of research which we may hope will be prosecuted in the near future.

The effect of pituitary preparations on kidney secretion has given rise to a considerable amount of controversy, the interest of which lies rather in the domain of pharmacology than within the scope of this discussion. It is mentioned here because some of those who have observed diuresis after the injection of pituitary extracts have indicated the possibility that the phenomenon is a consequence of increased arterial pressure. Others have recorded an antidiuretic effect. Both accounts are harmonised by some very important observations of Stehle (1926), who has not been content to measure the amount of fluid, but has made a careful study of the saline constituents of the urine after injection of pituitary preparations; and draws conclusions which do not support the pressor interpretation stated above. As moreover Stehle's observations suggest a possible hypothesis of the *modus operandi* of the pituitary autacoid in relation to the physiology of the cell itself, they call for special comment.

According to the observations of this worker, injection of pituitary extract in the dog produces an initial inhibition (anuria or oliguria) of the flow, followed from five to twenty minutes later by a definite diuresis. The urine collected during the latter period shows on analysis a much higher content of chlorides than does normal urine, a circumstance which in itself does not favour the view that the increased urinary flow is of vasomotor origin. What is perhaps more suggestive is the fact that along

Experiment 6*

Time	Vol. per min.	Cl per min.	Cl %	Na per min.	K per min.
2.12—3.02	0·147	0·138	0·094	0·182	0·131
3.04	1 c.c. of pituitrin injected.				
3.05—3.13	—	—	—	—	—
3.13—3.28	0·55	2·49	0·45	1·34	2·72
3.28—3.45	1·18	4·86	0·41	3·22	2·82
3.45—4.00	0·59	3·13	0·43	2·83	1·56

[* Stehle. *Am. Journ. Physiol.* vol. **79**, 1927.]

with the increased chloride concentration it is found that the potassium content of the urine is greater than the sodium content. The reverse is the case during the control period. So that we must regard the diuresis as a salt diuresis, and moreover, a salt diuresis which is brought about by the discharge of tissue salts, in particular potassium salts, from the tissues, there being no reason to locate the discharge specifically in the kidney itself. The increased output of chlorides is associated with a higher phosphate content, and we may thus infer a change in the permeability—in the broadest sense of the term—of the tissue cells or at least the cells of certain tissue, a change correlated with the removal of potassium. In view of the similarity of action of pituitary extracts and of potassium ions on uterine muscle melanophores and other effectors, taken in conjunction with Clark's observations showing that removal and addition of potassium ions have the same effect on mammalian plain muscle, Stehle's observations on the diuretic action of pituitary extracts provides a possible clue to the most fundamental and the most obscure aspect of endocrinology—the physical mechanism of the action of autacoids on the effector unit.

REFERENCES

ABEL. *Op. cit.* (chapter IV).

ATWELL and MARINUS. *Am. Journ. Physiol.* 47, 1918.

DALE and DUDLEY. *Journ. Pharm. Exp. Ther.* 18, 1921.

DREYER and CLARK. *Journ. Physiol.* 58, 1924.

DUDLEY. *Journ. Pharm. Exp. Ther.* 14, 1919; 21, 1923.

HERRING. *Quart. Journ. Exp. Physiol.* 6, 1913; 8, 1915.

HOGBEN and SCHLAPP. *Ibid.* 14, 1924.

HOGBEN, SCHLAPP and MACDONALD. *Ibid.* 14, 1924.

HOGBEN and DE BEER. *Ibid.* 15, 1925.

SCHLAPP. *Ibid.* 15, 1925.

STEHLE. *Am. Journ. Physiol.* 79, 1926.

Chapter VI

ENDOCRINE FACTORS IN METABOLISM

§ I

In the preceding chapters we have paid attention to the relation of internal secretions to effector response. In the two which follow the rôle of internal secretions in co-ordinating metabolic activities and developmental processes will be the theme for discussion. In both these fields there has been considerable progress during the last decade in the one case through investigations of a more or less clinical nature; in the field of developmental physiology through researches on the lower Vertebrates. The intervention of endocrine factors in normal metabolism will be considered under three headings: the relation of the thyroid to basal metabolism, the relation of the pancreas to carbohydrate metabolism, and the functional significance of the parathyroid glands.

The functional activity of the mammalian thyroid is of special interest partly on historical grounds as constituting the first case in which the administration of tissue extract was brought into harmonious relation with the effects of removal or disease of the organ, and partly, because with the exception of the chromaphil tissue it is the only gland from which a single chemical substance having all the physiological properties of its extract has as yet been isolated. An account of the early experiments of Horsley (1884), who described the effects of thyroid extirpation in monkeys, and those of Murray (1891), who showed that glycerol extract of fresh sheep's glands dissipated the symptoms associated with a diseased condition of the thyroid in the human subject, will be found in most treatises on endocrinology written from the clinical standpoint. As the early work was mainly concerned with the therapeutic and pathological aspects of the problem, and much confusion resulted from the difficulties of distinguishing between the effects of thyroidectomy and parathyroidectomy, it will be more profitable for our present object to confine the discussion to modern work which applies to Mammals in general, leaving out of account such issues as are of special interest to clinicians only.

Adhering, however, to historical precedent, the first question which arises is the nature of the effects which follow complete removal of the thyroid (without the parathyroid) glands in the Mammal. The visible manifestations, though in some points qualitatively similar, differ in young and old individuals and are not precisely identical in different species. On the other hand, a study of the metabolic process reveals a phenomenon common to all Mammals. This is the remarkable reduction in the basal metabolic rate, measured either as heat production or gaseous exchange, which is associated with removal or disease of the gland. Magnus-Levy (1897) first showed in a protracted series of observations on a patient suffering from the pathological condition known as myxoedema, that the standard metabolism associated with the disorder was much lower than is generally the case with normal persons, being 2·9 c.c. O_2 per minute per kilo body weight. After administration of thyroid substance in tablet form *per os* oxygen consumption rose during three weeks from this value to 5·5 c.c. per kilo per minute. On cessation of the treatment it fell to its former level, i.e. to 3·0 from seven to thirteen weeks later. The treatment was repeated twice with cessation and reappearance of the myxodematous symptoms. The effect of thyroid removal and thyroid extract upon standard metabolism appears to be a general one. In cats and rabbits the diminution begins about a week after operative removal of the gland and a constant low level is attained within three weeks or a fortnight. The reduction is of the order of 40 per cent. At this reduced rate it may remain for years (rabbits) unless accessory thyroid tissue has developed in the intervening time. The following protocol of a typical experiment by Marine and Baumann (1922) illustrates the effect of extirpation of the thyroid on the metabolic rate of a normal adult rabbit:

Date	Room temp. °C.	Weight in grm.	O_2 in grm. per 2 hours	Calories per kilo per hour
May 31	22·8	1827	2·885	2·77
June 2	23·1	1827	2·790	2·65
June 6	23·1	1806	2·690	2·63
June 9	24·1	1823	2·720	2·62
June 13	Thyroidectomy			
June 17	24·8	1915	2·115	1·93
June 21	25·8	1969	1·955	1·74
June 27	27·3	2047	2·240	1·92
July 5	26·9	2076	2·165	1·83

It would seem that the effects of administering thyroid extract by injection or feeding with glandular material are more pronounced in the case of thyroidectomised animals than with normal individuals. The ultimate nature of the metabolic changes is somewhat obscure. According to Hewitt (1915) administration of fresh ox thyroid to rats results in (1) a retention of nitrogen, and (2) an increase in the ammonia content of the urine. Cramer and Krause (1913) found that thyroid feeding was followed by complete disappearance of glycogen from the liver without glycosuria even when the diet was rich in carbohydrates. Cramer and M'Call (1917) bring forward evidence that the increased production of carbon dioxide and consumption of oxygen which accompanies the administration of thyroid substance is mainly due to the oxidation of carbohydrates in the tissues.

The manifest effects of thyroid removal and administration in Mammals depends to some extent on the age of the animal; and are partly concerned with growth. For reasons which will be mentioned later it seems desirable to refer to them at this stage rather than in our treatment of the relation of the thyroid to developmental processes. When the thyroid is removed without extirpation of all parathyroid tissue the most striking effect in young animals is a retardation of growth. Adult Herbivora (rabbit, sheep or goat) may show very little change other than a dryness and thickening of the skin, shedding of hair, increase of body weight and lowering of temperature. The cutaneous symptoms have been specially studied by Sutherland Simpson (1924) in the sheep. Removal of the thyroid always results in diminished fleece weight and in some cases a falling out of the wool. When the operation is performed on young animals at from three to four weeks from birth the operated animals are noticeably stunted in growth: the weight of the control may be three times that of its cretinous co-twin. If the operation is delayed till the third or fourth month the effects are only slight. While there is no doubt that removal of thyroid glands in young animals is followed by retardation of growth analogous to that which is characteristic of *cretinism* in the human subject, few authors have recorded symptoms identical with the cutaneous swellings of the abdomen and face in *myxoedema* after thyroidectomy in other Mammals. A discussion of this question will be found in Sutherland Simpson's paper.

A disorder corresponding in some respects to the pathological condition known as goitre has been recorded in Fish: and has been studied in the brook trout especially by Marine and Lenhart (1910–13). It is characterised by "excessive growth of thyroid tissue leading to the formation of visible tumours which may appear ventrally in the region below the base of the tongue or dorsally in the floor of the mouth and pharynx between the first and third gill arches." Fish suffering from hyperplasia of the thyroid gland are abnormally fat, weak, sluggish and more susceptible to disease. The administration of iodine compounds removes or checks the development of the disorder. Its occurrence illustrates the susceptibility of the thyroid to dietetic influences, a susceptibility well seen in Mammals. So far as is definitely known at present its appearance is confined to carnivorous Fish reared in captivity; and its spread has been parallel with the development of fish culture and artificial rearing of game Fish in large numbers. According to Marine it is non-infectious and non-contagious, being symptomatic of faulty nutrition. Feeding on an exclusive diet of liver is a major aetiological factor, at once corrected by changing to a diet of whole sea Fish. Water was found to play no essential part in the aetiology or transmission of the disease in the hatcheries studied by him.

The characteristic consequences of thyroid removal in Mammals, like the symptoms of thyroid degeneration in the human subject, can be prevented or removed by administration of the gland substance. Concerning the effect of thyroid extract or feeding with fresh glands on the growth rate of normal animals there is considerable disagreement. Relatively large quantities produce loss of weight in Mammals and Birds. In the latter case the treatment is associated with moulting (Crew) and in cocks, according to Torrey and Horning (1922), with the assumption of hen-feathering in the neck and saddle hackles and rectrices. It may be mentioned in this connexion that no undoubtedly successful attempts to remove the thyroid in Birds are on record. As in Mammals the organ readily regenerates if any glandular tissue is left behind. Some authors, among whom Hewitt (1915) may be mentioned, have found that small doses induce increase in body weight of normal rats. Hoskins (1916), who alone among workers on this problem has used genetically standardised material, obtained negative results.

At present the methods available for testing the physiological activity of thyroid preparations are all open to one objection or another for purposes of quantitative comparison. As a means of assay the effect upon the metabolic rate is obviously costly in materials, apparatus and time. Estimation of the iodine content is a useful guide, but its use as a criterion of activity implies the unproven assumption that all the iodine is present in active form. The action of thyroid extracts as a means of inducing metamorphosis in Amphibia is open to objections that will be discussed later. The acetonitril test of Reid Hunt (1905) is useful for detecting minute quantities. Its specificity depends upon the fact that thyroid administration increases the resistance of mice while lowering the resistance of other Mammals to acetonitril. Its use as a method of assay depends upon comparing the minimal lethal dose of the drug when administered with and without the thyroid preparation. This involves feeding the experimental animals from 2 to 14 days with the preparations, and using for both series mice of uniform stock, of approximately the same age and of the same sex fed on a uniform basal diet. For full details of the test, and discussion of its bearing on various problems connected with thyroid function, the recent paper of Reid Hunt (*vide infra*) may be consulted.

The isolation from the thyroid gland of a single crystalline containing iodine and having all the specific physiological attributes of thyroid extract was first accomplished ten years ago by Kendall. It had long been known that the thyroid of the Mammal contains a high percentage of iodine as compared with other animal tissues and from the recognition of this fact by Baumann (1895) numerous researches were directed to the separation of an active iodine compound from the gland. Oswald (1899) separated the proteins of the thyroid from a relatively stable fraction whose activity was proportional to its iodine content. Marine and his co-workers (1914–16) have shown that the perfused and isolated mammalian gland readily absorbs and stores iodine if the latter is present in the circulating fluid. A high percentage of iodine is shown by the analyses of Cameron (1914) to be present in the thyroid of all Craniates; and no other tissues contain iodine in quantities which are by comparison significant. The following list gives the percentage of iodine present in the thyroid of representatives of different classes:

Scyllum canicula	1·160	*Corvus*	0·751
Raia clavata	0·404	*Columba*	0·485
Rana pipiens	0·063	*Lepus*	0·067
Alligator	0·059	*Cavia*	0·114

Iodine is present often in relatively high percentage in the body of Ascidians but chiefly in the test. It is not present in appreciable quantity in the endostyle. To give a clearer appreciation of the significance of the figures quoted above, it may be mentioned that the highest percentage of iodine found in other living tissues of the vertebrate body is less than 0·001 per cent. (Cameron, Boucet). Iodine is present in the Mammal at an early stage (third to fourth month at least), according to Fenger in the foetal life of ruminants. Its presence in active form at this stage is shown by its power to induce metamorphosis in axolotls (Hogben and Crew). We may conclude from the analyses of Cameron and the structural similarity of the gland throughout the Craniata that the same active substance is produced in all Vertebrates. This raises the interesting question of the sources of iodine in the animal body. Appreciable quantities of this element are present in the bodies of sponges, corals, tubicolus annelids and ascidians. In marine algae the limits observed in reliable analyses are 0·001 to 0·7 per cent. The Laminarias contain amounts greater than 0·1 per cent.; and all the species with a high iodine content occur below the tidal zone and are never exposed. Terrestrial plants contain very much less iodine. But Cameron concludes from his studies that the ultimate source of iodine in animal tissues resides in the power of the plant cell to store it, and that differences in the iodine content of the gland can in general be correlated with difference of diet. The quantity of iodine present in sea water (McLendon) is 0·05 mg. per litre.

The chemical investigation of the thyroid before 1914 had resulted in the preparation of two types of active material, one, the *iodothyrin* of Baumann, a mixture of substances which could not be looked upon as a concentrated form of the desiccated glandular material, the other, *thyreoglobulin* of Oswald, was the product of removal of an inactive protein fraction. Kendall's investigations had their starting point in the discovery that the thyroid autacoid is uniquely stable to alkali hydrolysis. This resistance to hydrolysis in alkaline solution makes it possible to break down the thyroid proteins in a mixture of 90 per cent. alcohol and 2 per cent. sodium hydrate. On investigation of the acid solubility of the products of this hydrolysis it was found that the active portion was insoluble in acid. Further purification was carried out by precipitation of impurities in a solution of the acid-insoluble active fraction with barium hydroxide till a product with 60 per cent. of iodine was obtained in pure crystalline form. After the accumulation of 7 grm. of this substance Kendall (1917) made his first determination of the constitution. Thyroxine, as this substance was named by its discoverer, is a white, odourless, crystalline compound which is insoluble in any organic solvents except those of a decidedly basic or acid nature. It is amphoteric. It is not easily oxidisable, nor is it readily reduced. It is stable towards heat and has a melting point about 250° C. Upon exposure to sunlight it is however unstable in weakly alkaline solutions.

While credit is due to Kendall for having first isolated from the thyroid gland a single crystalline iodine compound which has been proved by numerous clinical investigations, such as those of Boothby and Sandiford (1924), to be a substance having the physiological properties of thyroid extracts, it is doubtful whether the views on the chemical constitution of thyroxine which Kendall has put forward are founded on sufficient evidence. In fact the much more detailed work of Harington (1926), who has given an entirely different empirical and structural formula for thyroxine, makes it unnecessary to set forth Kendall's views in this place.

In his experiments Kendall obtained a very small yield—about a gram and a half from two hundredweights of fresh glandular materials. Harington's method is more economical, the yield being about the same quantity of thyroxine for a kilo of dried gland. He substitutes for NaOH, hydrolysis with dilute baryta in the initial stage of the breaking down of the proteins. The detailed procedure advocated by Harrington is as follows. Half a kilo of desiccated gland is boiled for 6 hours under a reflux condenser with 10 per cent. baryta. After standing and filtering, the clear yellow filtrate is acidified with HCl till barely acid to Congo red, when a pale flocculent precipitate A settles. The material B from the previous filtration and the fraction A are both worked up. Precipitate A is redissolved in 250 c.c. of water with the aid of ammonia: crystalline $Ba(OH)_2$ is added in amount equivalent to 40 per cent. concentration, and the solution is heated for 18 hours on a steam bath, filtered hot through a Buchner funnel, the residue only being retained. This precipitate is suspended in 1 per cent. soda, boiled and treated with a strong solution of sodium sulphate in slight excess, then freed from $BaSO_4$ by filtration, and again boiled with addition of 50 per cent. sulphuric acid to make the reaction just acid to Congo red. A heavy granular precipitate which then forms is removed, redissolved in 20 c.c. normal soda, to which alcohol is subsequently added (up to 80 per cent.). After filtering once more and acidifying with acetic acid, crude thyroxine separates out. In working up material B, the precipitates from the preliminary hydrolysis of the gland are ground up and boiled with two litres of 2 per cent. soda, slight excess of sodium sulphate added, the filtered solution treated to acidify with hydrochloric acid, and then dealt with in the same manner as A. The crude thyroxine is equivalent to 0·125 per cent. of fresh gland substance. It is purified by repeated solution in alkaline alcohol and precipitation with acetic acid. It then forms rosette-shaped sheaves of fine needle-like crystals, which on heating darken at 220° C. and melt with evolution of iodine at 231–233° C. In solution it is optically inactive, being racemised by previous treatment with alkali.

Kendall regarded thyroxine as 4-, 5-, 6-trihydro-, 4-, 5-, 6-triodo-2-oxyindol-propionic acid and gave to it the empirical formula $C_{11}H_{10}O_3NI_3$. Harington on the other hand regards it as $C_{15}H_{11}O_4NI_4$, which agrees very well with Kendall's as well as his own data. His view of the constitution is entirely different from Kendall's. According to Harington it is a tetraiodo substituted derivative of β-[4-(4'-hydroxyphenoxy)-phenyl]-α-amino-propionic acid, the p-hydroxy-phenyl ether

of tyrosine, with the iodine atoms occupying probably the
3, 3', 5, 5' positions. Both regard it as an iodine derivative
of an amino acid containing a benzene ring. Harington's formula
indicates that there are two benzene rings in the molecule.
As Harington's publication did not appear until the present
chapter was in the press, it is impossible to give an adequate
discussion of his experiments; the more extensive character of
his investigation however makes it unnecessary to deal in detail
with the earlier work of Kendall. The formula derived by Harington is based on an exhaustive study and successful synthesis
of the reactions of the deiodised compound desiodothyroxine,
formed by splitting off the iodine quantitatively by reducing
agents with substitution of an equivalent quantity of hydrogen
(4 mols per mol of thyroxine). Thus the constitution attributed
to thyroxine by Harington is based on the assumption that the
iodine is removed from thyroxine without induction of secondary
reductive changes in the molecule. The essential stages in the
synthesis of this compound are thus epitomised by the author
himself:

"On condensing p-bromoanisol by the method of Ullman and
Stein with the potassium salt of p-cresol in presence of copper
bronze, there was obtained 4-(4'-methoxyphenoxy)toluene:
this compound on boiling with hydriodic acid gave 4-(4'-hy-
droxyphenoxy)toluene, which proved to be identical with the
substance $C_{13}H_{12}O_2$ which resulted from the mild potash fusion
of desiodothyroxine. Further, on boiling with permanganate the
4-(4'-methoxyphenoxy)toluene gave an acid 4-(4'-methoxy-
phenoxy)benzoic acid which was identical with the acid $C_{14}H_{12}O_4$
obtained as the final oxidation product of desiodothyroxine after
methylation. p-bromoanisol was then condensed similarly with
potassium phenate, giving (4'-methoxyphenoxy)benzene: from
this, by Gattermann's hydrocyanic method, was prepared an
aldehyde. This aldehyde was identical with the aldehyde $C_{14}H_{12}O_3$
obtained in the degradation and was further proved to be 4-(4'-
methoxyphenoxy)benzaldehyde by the fact that it gave an acid
identical with 4-(4'-methoxyphenoxy)benzoic acid....Desiodo-
thyroxine was synthesised by two methods: (a) by the method
of Sasaki (1921), 4-(4'-methoxyphenoxy)benzaldehyde was con-

densed with glycine anhydride in presence of acetic anhydride
and sodium acetate; on boiling the condensation product with
hydriodic acid and red phosphorus it underwent simultaneous
reduction, hydrolysis and methylation with the formation of
β-[4'-(4'-hydroxyphenoxy)phenyl]-α-amino-propionic acid: (b)
by the method of Wheeler and Hoffmann (1911), 4-(4'-
methoxyphenoxy)benzaldehyde was condensed with hydantoin and
the condensation product boiled with hydriodic acid and red
phosphorus in the same way. In both cases the synthetic amino-
acid was identical with the natural desiodothyroxine...."

The constitutional formula of desiodothyroxine is thus:

$$HO \langle \bigcirc \rangle - O - \langle \bigcirc \rangle CH_2 \cdot CH \cdot NH_2 \cdot COOH.$$

Direct evidence that thyroxine is liberated into the blood stream
is still lacking. And there are still insufficient grounds to substantiate this conclusion through indirect channels, since the seat
of the oxidative changes has not yet been investigated. Circumstantial evidence points to the likelihood that thyroxine facilitates
cellular oxidations. And it is regrettable that the action of thyroxine
or thyroid extract on oxygen consumption in isolated tissues has
not been investigated, though with the aid of Barcroft's manometer
the problem is not a formidable one. Carell (1913) has recorded
the observation that growth of surviving tissues in artificial culture
media is expedited by the presence of thyroid extract and other
observers have described increased cell division in Protozoa with
similar treatment. This is, however, denied by Woodruff and
Swingle (1924).

REFERENCES

BOOTHBY and SANDIFORD. *Physiol. Review*, **4**, 1924.
CAMERON. *Journ. Biol. Chem.* **18–23**, 1914–15.
HARINGTON. *Biochem. Journ.* **20**, 1926.
HEWITT. *Quart. Journ. Exp. Physiol.* **8**, 1914–15.
HOSKINS. *Journ. Exp. Zool.* **21**, 1916.
KENDALL. *Journ. Biol. Chem.* **39**, 1919; *Endocrin.* **3**, 1919.
KENDALL and OSTERBERG. *Journ. Biol. Chem.* **40**, 1919.

Since the above was written a further contribution of Harrington and Barger
announces the successful synthesis of thyroxine itself. (See *Biochem. Journ.* **21**,
1927.)

MARINE and LENHART. *Journ. Exp. Med.* 12–17, 1910–13.
REID HUNT. *Am. Journ. Physiol.* 63, 1923.
SUTHERLAND SIMPSON. *Quart. Journ. Exp. Physiol.* 14, 1924.
TORREY and HORNING. *Proc. Soc. Exp. Biol. Med.* 19, 1922.
WOODRUFF and SWINGLE. *Am. Journ. Physiol.* 69, 1924.

§ II

The relation of the pancreas to carbohydrate metabolism was first recognised through experiments of Mering and Minkowski (1899) demonstrating the occurrence of severe and fatal diabetes after removal of the pancreas in dogs. This observation was abundantly confirmed by other investigators. Sscobolew (1902) found that when degenerative atrophy of the pancreatic acini was produced by ligature of the ducts, no glycosuria ensued. MacCallum (1909) extended and confirmed the results of Mering and Minkowski and of Sscobolew. The significance of their somewhat divergent findings became clearer when Kirkbride (1912) showed that the atrophic pancreas contained healthy groups of epithelial elements of the same type as those already described in the normal gland as "islets of Langerhans." Clarke (1916) found that the sugar consumption of the mammalian heart is increased when the perfusion fluid is first passed through blood vessels of the pancreas, and that sugar passed through blood vessels of the pancreas undergoes stereoisomeric changes though its reducing properties are unaffected. Of early attempts to prepare an active pancreatic extract, however, some reference is due to the work of Zuelzer (1908), who showed that an alcoholic extract of pancreas antagonised the glycosuria resulting from injection of adrenaline, but the harmful nature of his preparations prevented a conclusive demonstration of the existence in the pancreas of a substance which lowers blood sugar. Various other attempts were made to prepare from pancreatic tissue an extract which would rectify the disturbances in carbohydrate metabolism resulting from pancreatectomy. But until 1922 efforts directed to this end were practically fruitless.

It has already been mentioned that secretin and also the pituitary autacoids are readily destroyed by trypsin. In reviewing the literature dealing with the relation of the pancreas to diabetes Banting (1920) recognised that the presence of trypsin in the secretory tubules might be a significant factor in the failure of previous attempts to obtain extracts of the glands which would influence carbohydrate metabolism; and put forward the suggestion that advantage might be taken of the degenerative changes which occur in the acini after ligature of the ducts to prepare an active extract. Two years later Banting and Best (1922) published from Macleod's laboratory an account of experiments in which this possibility was first put to experimental test. They showed that saline extracts of the " degenerated " dog's pancreas removed from seven to ten hours after ligature of the ducts when injected into the same animal intravenously invariably exercised a reducing influence on the sugar content of the blood and the amount of sugar excreted in the urine of diabetic dogs: that the extent and duration of the reactions varied directly with the amount of extract injected; and that the extract was inactivated by pancreatic juice. Their experiments proved that the active substance was destroyed by boiling; and that injection of their extracts enabled a diabetic animal to retain a much higher percentage of sugar than it would otherwise tolerate. They also showed that the action was specific and was not due simply to dilution of the blood, since there was no appreciable lowering of the blood sugar when large quantities of saline or other tissue extracts were injected.

Observations indicating that the presence of enzymes in the mammalian pancreas cannot be detected in the foetus until the later stages of gestation and that pancreatectomy was not followed by severe glycosuria until the young were born, suggested a further means of preparing an extract rich in the anti-diabetic substance and free from the destructive enzymes present in the pancreatic juice. This method was employed by the same authors shortly after the completion of the experiments described above. Intravenous and subcutaneous injections of neutral saline extracts of the ox foetus at about the fifth month had the same effect in lowering the blood sugar as extracts of the degenerated pancreas of the adult; and daily injections of pancreatic extract were used to sustain life in a dog for seventy days after total removal of the gland. Two new observations on the chemical properties of the active substance, its failure to dialyse through parchment and its destruction at 65° C. were recorded at this stage.

These initial experiments were carried out with dogs in which glycosuria had been induced by pancrcatectomy. The efficacy of alcoholic extracts of pancreas ("insulin") to lower the blood sugar of normal animals was also established in the Toronto laboratory, and a crude method of assay thus became possible. When the extract is injected into a normal rabbit violent convulsions occur when the blood sugar falls below about 0·045 per cent. As a unit for quantitative treatment the number of c.c. of extract required to bring down the concentration of blood sugar within four hours to the level at which convulsions occur was proposed. This is usually referred to as a "rabbit unit." The animal may recover from these early symptoms, but with active preparations the hyper-excitability usually becomes acute, convulsive seizure involving the whole body result, and the animal lies in the interval in a more or less comatose condition on its side. When the fall in blood sugar is well established, piqure, injection of adrenaline, asphyxia and ether anaesthesia do not bring about marked hyperglycaemia. An important advance was made when Collip (1922) elaborated a simple method of obtaining an extract from normal pancreatic tissue. Freshly minced glands were added to an equal volume of 95 per cent. alcohol, the mixture being allowed to stand for a few hours, and filtered. The filtrate was mixed with twice its volume of 95 per cent. ethyl alcohol to precipitate the major part of the proteins which after a few hours' standing were filtered off. By evaporation *in vacuo* or in a warm air current the lipoids were then extracted with ether, all the active substance remaining in the alcoholic layer. The latter was mixed with absolute alcohol to precipitate the active substance (along with various adherent substances of unknown composition), and the final precipitate was extracted with water. This extract contains no trypsin and very little protein matter, it is lipoid free and can be injected subcutaneously without local reactions[1].

It is from the comparative physiology that we derive the clearest evidence as to the localisation of the endocrine function of the

[1] *While this book was in the press Abel announced the isolation of a pure crystalline substance with the properties of insulin.*

pancreas. As already stated, there are present in the pancreas, in addition to the cells of the secreting tubules, groups of epithelial elements which are devoid of lumina commonly referred to as the "islets of Langerhans." Schafer was the first to suggest that the islets of Langerhans are responsible for the specific relation of the pancreas to carbohydrate metabolism; and the same author suggested the name *insulin* for this active constituent, at this time a purely hypothetical entity. The view that these structures are morphologically and ontogenetically independent of the undoubtedly exocrine portion of the gland was for some years subjected to considerable criticism. But it now appears from the recent work of Macleod to rest on a secure foundation.

Rennie (1904–5) first drew attention to the presence of islet tissue in the Teleostei and discovered that the islets in these fishes are collected into nodules which are often encapsulated and thereby separated from the zymogeneous tissue. In the cartilaginous fishes the islets and zymogenous tissues on the other hand have the general relations characteristic of Vertebrates. Rennie himself was not able to obtain experimental proof that the isolated islet tissue of bony fishes contains a substance which when injected into the circulation lowers the blood sugar or ameliorates diabetic symptoms. By alcoholic extraction of the whole pancreas Macleod (1922) obtained insulin from the skate and the dogfish. The following protocol of one of Macleod's experiments illustrates the methods employed. The glands were removed on different days from 43 small skate and placed in alcohol in the refrigerator, ground with sand and extracted with acid alcohol, filtered, and later evaporated in warm air. The final extract of 22 c.c. corresponded to 120 grm. of pancreatic tissue. 10 c.c. was injected into a rabbit (1·6 kg. body weight):

9.45 a.m.	0·120 per cent. blood sugar
9.55 ,,	Injection
10.55 ,,	Violent convulsions of the usual type
11.0 ,,	Partial recovery
11.02 ,,	0·035 per cent. blood sugar
12.0 noon.	Violent convulsions which ceased when 5 grm. of dextrine in solution was injected subcutaneously
3.0 p.m.	0·064 per cent. blood sugar
9.0 a.m.	Dead

In experiments on Teleostei (*Myoxocephalus, Anguilla, Lophius, Zoarces*) it was found that strong extracts of pancreatic tissue devoid of islet nodules had no effect on the blood sugar, whereas portions containing islet tissue or islet capsules alone yielded highly potent extracts. The following protocol illustrates the nature of these experiments. The principal islets were removed from 35 specimens of *Myoxocephalus* and placed in dilute acid alcohol. After passing through muslin there was 1·2 grm. of islet substance. The final extract after removal of alcohol measured 5 c.c., of which 2 c.c. was injected into a rabbit (2·4 kilos body weight).

10.0 a.m.	0·114 per cent. blood sugar
	Injection
11.20 ,,	0·020 per cent. blood sugar
11.35 ,,	Convulsions
11.45 ,,	0·020–0·026 per cent. blood sugar
11.50 ,,	Convulsions less severe
12.05 p.m.	0·04 per cent. blood sugar
12.10 ,,	Severe convulsions, 4 grm. dextrose injected subcutaneously
2.30 ,,	0·090 per cent. blood sugar, rabbit normal
5.40 ,,	0·062 per cent. blood sugar
6.00 ,,	More dextrose injected
	Next morning the rabbit was normal.

Macleod concludes that these results afford strong direct evidence for Schafer's hypothesis that insulin, as its name implies, is derived from the insular and not the zymogeneous tissue of the pancreas. Concerning the significance of the islet tissue in the metabolism of the lower Vertebrates there are few important data other than observations of McCormick and Macleod (1925) showing that isletectomy in fishes is followed by hyperglycaemia. Huxley and Fulton (1924) noted convulsions after injection of insulin in tadpoles; but Olmstead (1924) recorded analogous symptoms both in frogs and catfishes in which no lowering of the blood sugar could be detected. Noble and Macleod (1923) could not detect any action of insulin on the blood sugar of the tortoise.

As regards Birds, Cassidy, Dworkin and Finney (1926) have shown that the fall of blood sugar content after injection of insulin in the domestic fowl is relatively greater than in Mammals: the lowering of the blood sugar is accompanied by a fall of body temperature; but the convulsions so characteristic of the effect of injecting insulin in the Mammal are not found to occur.

There are now available a number of methods for obtaining insulin from the pancreas; Seeing that the active substance is destroyed by proteolytic enzymes, the extraction must be carried out in such a way that the trypsin is not allowed to act upon it. This may be done by ensuring the absence of trypsin at the outset by producing degenerative changes in the acini, extracting from the foetal glands before the endocrine portion contains any trypsin, or making the extract from islet tissue alone, when as in some bony fishes the latter occur in separate masses. In practice it is customary to destroy the enzyme. Collip's method has already been mentioned. Shortly afterwards Doisy, Sornogyi and Shaffer (1923) obtained a better yield by extracting in strongly acidified alcohol (1200 c.c. of 95 per cent. alcohol, 40 c.c. concentrated HCl, 300 c.c. water, 1 *kilo* pancreas). The best yield at present is obtained by the more recent method of Dudley and Starling (1924). The process depends upon the fact that insulin is more readily absorbed from acid solution, but, like Collip's original method, makes use of the fact that insulin though soluble in 80 per cent. alcohol is precipitated by a concentration of about 93 per cent. The details are as follows:

One kilogram of fresh ox pancreas is minced into one litre of 95 per cent. spirit; 85 grm. sodium bicarbonate is stirred into the mixture, which is then poured into the mincer and re-minced; this procedure is repeated a second time. The mixture is then allowed to stand with frequent stirring for two hours at room temperature, or if desired may be kept in the cold room overnight. It is then poured on to a stout cloth, which is squeezed as thoroughly as possible in a suitable press. The turbid filtrate, which has a pH of about 7·5, is treated with 1½ times its volume of 95 per cent. spirit and placed in the cold room (- 3° C.) overnight. It is then filtered through paper, and 10 c.c. of glacial acetic acid added to the filtrate, which is then evaporated *in vacuo* in a water-bath at 40–45°. When the volume of the residue has reached about 150 c.c., the fat which has separated is removed by shaking out with light petroleum and the aqueous layer is freed from traces of this solvent by a few minutes' distillation under the conditions described above.

Four volumes of absolute alcohol are added to the aqueous solution, and after standing for about ten hours in the cold room the alcoholic solution is carefully decanted from the precipitate which has formed, and two volumes of absolute alcohol are added to it. After standing in the cold room for 12–24 hours, the supernatant liquid is carefully decanted and the precipitate is washed into centrifuge tubes with absolute alcohol and dry ether, and finally dried in a vacuum desiccator over sulphuric acid.

By this method 2 grm. of a white hygroscopic powder may be obtained from a kilo of pancreatic tissue. This quantity of "crude insulin" represents about 412 rabbit units. It contains over 50 per cent. of inorganic salts (phosphates, carbonates, etc.), gives the tests for tyrosine, tryptophane and laevulose, and contains a depressor substance (or substances). A dilute (1·5 per cent.) solution at a pH about 5·0 is precipitated by Dudley with half a volume of saturated aqueous picric acid. The picrate is collected by centrifuging, dissolved in sodium carbonate and re-precipitated. The moist precipitate is finally decomposed with acid alcohol. The precipitate formed in excess of acetone is at least ten times as concentrated as the crude product, 0·02 to 1 mg. being equivalent to a rabbit unit.

Dudley's product is not a pure substance. It gives a strong biuret action, a positive Pauly reaction for histidine and contains organic sulphur. It no longer gives the reactions for tryptophane, phosphorus or fructose. Its heat stability is greater than that of the original preparation of Banting and Best. In strongly alkaline solutions (decinormal soda) it is destroyed at body temperature within 1½ hours. It is readily destroyed by boiling in neutral or alkaline solution, though (like trypsin) much more stable to heat in an acid medium. It dialyses very slowly, if at all, through parchment, and does not pass through collodion. It is adsorbed from acid solutions by kaolin and animal charcoal. It is precipitated by picric, phosphotungstic, and trichloracetic acids, by uranyl acetate and by half saturation with ammonium sulphate and sodium chloride. In addition, it is destroyed by pepsin as well as trypsin. In short, all the available indications point to the view that insulin is a complex structure probably of protein-like nature. The discovery of insulin has given a new impetus to investigations in carbohydrate metabolism which will in all probability enrich physiological knowledge to a far greater extent than the mere addition of a new illustration of the rôle of internal secretion in the regulative processes of the organism. However it is with the latter issue that we are here concerned.

While there is little doubt that the islet tissue is particularly rich in insulin content, it would be wrong to regard insulin as a specific property of the pancreas in the sense that adrenaline is a specific product of the chromaphil tissues. Quite a considerable yield was obtained from salivary glands, kidneys and spleen by Baker, Dickens and Dodds (1924). Quite large quantities moreover were extracted from the tissues of a person who had died of diabetic coma. Moreover Best, Smith and Scott (1924) found a very appreciable amount of insulin even in the muscles of a depancreatised dog.

We have now to ask in conclusion whether insulin is actually set free into the blood stream. We know that it is possible to obtain from crushed yeast cells an enzyme which acts upon sugar in a manner precisely analogous to the fermentation produced by the intact organism. And the extraction of insulin from the pancreas does not of itself throw any light on the seat of the changes which are initiated by its active constituent. We can obtain an answer to the issue which has been stated above either by (a) seeking direct evidence of the liberation of insulin into the blood or (b) locating the seat of action of insulin in relation to carbohydrate metabolism.

As regards the first the experiments of Clough, Allen and Murlin (1923) show that it is possible to obtain by perfusion of the intact pancreas a high yield of the active substance. This strongly suggests that insulin is set free into the blood stream. But the clearest evidence for this conclusion lies in the localisation of the seat of action of insulin. Hepburn and Latchford (1922) have investigated the effect of insulin on the consumption of sugar by the excised mammalian heart. They found, like previous workers who have used Locke's method, an average consumption of 0·9 mg. of glucose per hour per gram of rabbit heart. On adding insulin to the perfusion fluid the average rose to 3·0 mg. per gram per hour, without manifestly affecting the cardiac rhythm. Further, it appears quite clear that insulin lowers the blood sugar by facilitating the uptake of sugar by the tissues; that is to say, that insulin is a freely circulating hormone.

Perhaps the most interesting observations on insulin from the standpoint of comparative physiology are those which have been made by Cassidy, Dworkin and Finney (1926). These workers have investigated the relation of sugar metabolism to the phenomenon of hibernation in Mammals. It has long been known that hibernation involves a pronounced lowering of body temperature, and of the respiratory quotient. Pembrey (1903) showed that in the dormouse the carbon dioxide production may be reduced to a hundredth of that of the normal animal when awake: temperature of the body sinks during hibernation from 30° C. to 12° C., rising as rapidly as 19° C. within an hour on waking, while the respiratory quotient is maintained during hibernation at the abnormally low value of 0·23. In the case of the torpid hedgehog, body temperature sinks to 16° C., the carbon dioxide production to a tenth of its normal value, and the respiratory quotient to as low as 0·51. In both cases the respiratory quotient rises rapidly to its normal value of 0·75–0·78 when the animal is awakened. The investigators referred to above have shown that lowering of the blood sugar by means of insulin abolishes the shivering reflex, when cats and dogs are subjected to cold; and lowering of the body temperature itself abolishes the characteristic convulsions following injection of insulin. The combined action of insulin and cold thus produces a state simulating in some respects hibernation. These experiments open up suggestive lines of enquiry into the mechanism of hibernation and the broader issue of the relation between body temperature and normal metabolism.

REFERENCES

BAKER, DICKENS and DODD. *Brit. Journ. Exp. Pathol.* 5, 1924.

BANTING and BEST. *Journ. Lab. and Clin. Med.* 7, 1922.

BANTING, BEST, COLLIP, MACLEOD and NOBLE. *Am. Journ. Physiol.* 62, 1922.

BEST, SMITH and SCOTT. *Am. Journ. Physiol.* 68, 1924.

CASSIDY, DWORKIN and FINNEY. *Am. Journ. Physiol.* 73, 1925, 75, 1926, and 77, 1926.

CLOUGH, ALLEN and MURLING. *Quart. Journ. Exp. Physiol.* (*Supplement XIth Internat. Congress*), 1923.

COLLIP. *Journ. Biol. Chem.* 55, 1923.

DUDLEY and STARLING. *Biochem. Journ.* 18, 1924.

HEPBURN and LATCHFORD. *Am. Journ. Physiol.* 62, 1922.

HUXLEY and FULTON. *Nature*, 113, 1924.

MACLEOD. *Journ. Metabol. Research*, 2, 1922.

McCORMICK and MACLEOD. *Proc. Roy. Soc.* B, 98, 1925.

NOBLE and MACLEOD. *Journ. Physiol.* 58, 1923.

OLMSTEAD. *Am. Journ. Physiol.* 69, 1924.

§ III

Several important communications dealing with the parathyroid glands have appeared during the past three or four years. Such information as we have at our disposal is almost exclusively drawn from experiments on Mammals and very largely from clinical observations. The subject therefore need only be touched on briefly in so far as it suggests new methods of enquiry into the general physiology of metabolism. From a variety of conflicting testimony with regard to growth, sexual development, blood sugar, etc., the more significant issues may be epitomised as follows.

A condition of tetany (muscular twitchings, etc.) is associated with atrophy or degenerative changes in the human parathyroids, and with experimental extirpation of the gland in other Mammals. In Carnivora, such as the dog, with very rare exceptions, tetany, which as a rule results in death, follows complete parathyroid-ectomy; but according to Sutherland Simpson (1924) the operation is tolerated much better in sheep. Concerning the significance of tetany two rival hypotheses—which are not necessarily incompatible—have been advanced.

In 1912 Koch claimed to have isolated guanidine derivatives from the urine of parathyroidectomised dogs: Koch's work was succeeded by a series of investigations by Paton and his collaborators, pointing to the conclusion that tetany is due to methyl guanidine intoxication, and that the function of the parathyroids is to remove from the blood a toxic substance. Paton and Finlay (1917) drew attention to the production of symptoms similar to the condition of tetany, when guanidine or methyl guanidine are introduced into the circulation. Burns and Sharpe (1917) in Paton's laboratory recorded a decided increase in the amount of guanidine and methyl guanidine in the blood of parathyroidectomised dogs and individuals suffering from idiopathic tetany. Later, Vines (1920) detected in parathyroid extracts an enzyme which effects the breakdown of guanidine *in vitro*. Nevertheless the work of the Glasgow school has not been accepted without criticism. The critical issue is whether guanidine derivatives are present in greater amount in the blood or urine of parathyroidectomised animals or cases of *tetania parathyreopriva*. Greenwald (1924) points out that in many of the experiments carried out with reference to this point a method involving precipitation with mercuric chloride and sodium acetate (or mercuric acetate) was employed; and this procedure involves oxidation of creatine into methyl guanidine glyoxilic acid if creatine is present. By a specially devised technique this author could not confirm Koch's observations, or the existence of a toxic substance in serum.

MacCallum, on the other hand, has called attention to the beneficial results of calcium administration after parathyroidectomy and to the fact that tetanic symptoms are also characteristic of calcium deficiency. Of these facts there now seems to be no doubt. Luckhard and Goldberg (1923) found that the tetanic symptoms and fatal consequences of parathyroid removal in dogs can be prevented in dogs fed on a meat diet, if 1·5 grm. of calcium lactate per kilo body weight is given every day. Further, the observations of Cruickshank (1923) and of Salvesen (1923) show that parathyroidectomy is followed by a fall in the blood calcium content. A new phase in the study of the parathyroids has now been initiated by Collip's announcement of a method of obtaining an extract which has a very definite effect upon the calcium content of the blood in normal and thyroidectomised dogs. The method of extraction is as follows:

Parathyroid glands were removed from oxen...and were at once placed in an ice box at − 4° C. When required for use a few glands were placed in a large Pyrex test-tube and covered with an equal volume of 5 per cent. HCl. The test-tube was placed in a boiling water-bath for 1 hour, the glands being broken up after a few minutes' heating by means of a glass rod. After the necessary period of digestion had elapsed the fat which had separated out as an oily layer on the surface of the extract was removed mechanically. The extract was chilled at once to room temperature and made alkaline to pH 8 by NaOH. HCl was then added slowly till a maximum precipitation of protein and protein derivatives occurred. The precipitation was removed...redissolved in weak alkali, and a second isoelectric precipitation carried out. The filtrates, or, if the centrifuge was used, the supernatant fluids, were combined, and the preparation represented an aqueous solution of the principle.

Injection of this extract raises the blood calcium of both normal and parathyroidectomised dogs. Usually an initial dose produces a definite effect from five to nine hours after: then follows return to normal. The initial rise in calcium content of the serum is greater in parathyroidectomised animals (Collip, Clark and Scott). Successive small doses always evoke a response in the blood calcium content of normal dogs and if continued the hypercalcaemia results in death. Successive injections into parathyroidectomised dogs not only restore the calcium content of the blood to a more normal level but prevent the appearance of tetanic symptoms. The phosphorus content of the blood rises during hypercalcaemia produced by administration of Collip's extracts. The reverse is the case after parathyroidectomy (Greenwald). The active substance is destroyed by proteoclastic enzymes. There are no definite data at present on which to base the conclusion that it is liberated into the circulation. Of the rôle of the parathyroids in the lower Vertebrates we know nothing at present.

Striking as are the data which emerge from these researches, a note of caution must be sounded. Recent work by Davies, Dickens and Dodds (1926) has raised the question as to whether the action of Collip's extracts is a specific property of the parathyroid glands. "Insulin," according to these workers, and also pituitary extracts, when injected into rabbits raise the serum calcium content considerably. Morever extracts of pancreas prepared in the manner prescribed by Collip for the preparation of the parathyroid autacoid, while manifesting a potent activity in relation to the serum calcium have no action on the blood sugar. At the moment more quantitative biochemical data are needed before it is possible to estimate the full significance of recent work on the parathyroid.

REFERENCES

BURNS and SHARPE. *Quart. Journ. Exp. Physiol.* 10, 1917.
COLLIP. *Journ. Biol. Chem.* 63, 1925.
CRUICKSHANK. *Biochem. Journ.* 17, 1923.
DAVIES, DICKENS and DODDS. *Biochem. Journ.* 20, 1926.
GREENWALD. *Journ. Biol. Chem.* 59–61, 1924.
PATON and FINDLAY. *Quart. Journ. Exp. Physiol.* 10, 1917.
SALVESEN. *Journ. Biol. Chem.* 56, 1923.
VINES. *The Parathyroids in disease.*

Chapter VII

THE RÔLE OF THE DUCTLESS GLANDS IN DEVELOPMENTAL PROCESSES

§ I

In concluding our survey of the rôle of endocrine agencies we shall now turn from the properties of the living organism as a finished product to the building up of a new animate unit. The fertilised egg bears within it the power to develop into an individual resembling the parents from which the sperm and egg were derived. Fertilisation starts in the egg a period of active cell division. In the initial stages of cleavage all the cells may be, and often are, for a considerable period, very much alike. As they go on dividing they differentiate progressively to build up the structural architecture of the new individual. In the early stages there is no increase in size; at some point, however, the developing organism begins to augment in weight and volume. This process usually goes on long after the final morphological order characteristic of the individual is completely established. Developmental phenomena may thus be considered under two headings: individuation and growth. Individuation, or the differentiation of structural pattern in cellular animals raises perhaps the most recondite issues in the whole field of biological enquiry. It is convenient to consider it separately in its spatial and chronological aspects, that is to say (i) the agencies which determine whether a particular region shall differentiate into one type of structure rather than another; and (ii) the agencies which determine the orderly sequence in which the differentiation of one structure follows another. It is with the latter only that we need here concern ourselves. The key to this problem is most readily accessible in those forms in which there is a phase of active growth intercalated before the final phase of individuation (metamorphosis). And the nature of the issue is well illustrated in Uhlenhuth's work on the transplantation of larval organs in Urodeles. When larval skin from *Amblystoma opacum* or *tigrinum* is grafted on to another individual the grafts develop the characteristic yellow spots, not simultaneously with one another as they would have done if left with the animal from which they were taken, but simultaneously with the development of the adult skin characteristic of the host: they undergo transformation only, if and when metamorphosis takes place in the host. When the graft is interspecific the same rule holds but the type of yellow spot developed in the grafted area is that of the species from which the skin was taken and not that of the host. Thus it is evident that in the development of the adult pigmentary characteristic there are two distinguishable forces at work, one which is specific and responsible for the kind of yellow spots which are developed, and one which is non-specific, not contained in the skin but produced in some other organ and necessary to initiate the developmental changes from the larval to the adult organisation. This "metamorphosis factor," we shall now proceed to investigate.

In no group of Craniates is the phenomenon of metamorphosis more widespread and more amenable to experimental treatment than in Amphibia. The nature of the developmental events which constitute "metamorphosis" in the tailless form (Anura) is not precisely the same as in the caudate species (Urodela), though in both cases it involves the transformation from an aquatic organisation to a terrestrial form. In Anura (frogs, toads) the hind limbs are quite minute until the onset of metamorphosis, and the fore limb rudiments do not become visible until transformation sets in. The outstanding changes which affect the bodily form are: (i) growth of hind limbs and appearance of the fore limbs, (ii) closure of the gill clefts; (iii) resorption of the tail. In Urodeles (salamanders, newts) both pairs of limbs are fully developed in the larval form; the external gills which are ephemeral structures in the anuran tadpole persist throughout larval life, and the tail is retained in the adult. Metamorphosis is signalised by the resorption of the external gills and shedding of larval skin. This is followed in many cases by the disappearance of the dorsal skin and by the complete closure of the gill slits. In both cases there are other less noticeable external features, protrusion of the eyes, development of eyelids and pronounced changes in the nature of the integument. As regards internal organisation the development of the lungs and gonads is independent of metamorphosis. Extensive ossification (as opposed to calcification which may take place at any time) is associated with the external changes.

§ II

In the analysis of this "metamorphosis factor" the earliest significant observation was that of Babak (1912), who first showed the efficacy of a diet of ox thyroid to induce metamorphosis in the persistently neotenous larvae of the Mexican salamander (cf. Fig. 20, Chapter III and Fig. 34). This fact, which will be dealt with later, did not attract much notice at the time. Shortly after, Adler developed the technique for experimental extirpation of ductless glands in larval Amphibia; and Gudernatsch (1912–14) began his work on the effects of thyroid feeding. These researches, initiated in Germany on the eve of the late war, represent the foundations of an immense and profitable superstructure for the later development of which the American school of experimental zoologists have been mainly responsible.

It was the experiments of Gudernatsch which first stimulated interest in the rôle of the thyroid. Gudernatsch investigated in large numbers of anuran tadpoles of two species, *Rana esculenta* and *R. temporaria*, the effect of a diet of various glandular and non-glandular tissues—thyroid, thymus, liver, muscle, suprarenal cortex and medulla, testis, ovary and pituitary from various Mammals (pig, rabbit, horse, dog, ox, cat). We need not here concern ourselves with the minor differences in regard to pigmentation, etc. The one unquestionable fact that emerged was that a diet of thyroid administered at any age after the tadpole was capable of taking solid food led within a few days to the onset of metamorphic symptoms. The time at which the feeding began was of no importance to the result. If the treatment began early the tadpoles fed on a thyroid diet underwent precocious metamorphosis with remarkable uniformity weeks before the controls fed on other tissues. The individuals of a thyroid-fed group in all cases transformed within about a day of one another. No such uniformity is seen in normal tadpoles; nor was it observed in the tadpoles fed by Gudernatsch on other tissues. Gudernatsch himself realised the very far-reaching nature of these experiments which were "to open a new and extensive field of work in experimental morphology." His results were soon confirmed by a number of workers—Lenhart (1915), Morse (1914), Barthelemez (1915), Lim and Swingle, to mention but a few names. Most spectacular of these researches were those of Swingle (1918) on the American bull frog, *Rana cates-biana*. In a state of nature it requires two seasons and sometimes three for the tadpoles of this species to attain the adult condition; and the gonads contain ripe germ cells long before the onset of metamorphosis. By feeding with desiccated mammalian thyroid Swingle was able to induce metamorphosis within three weeks from hatching, thus producing miniature frogs having all the bodily characteristics of the adult but larval gonads.

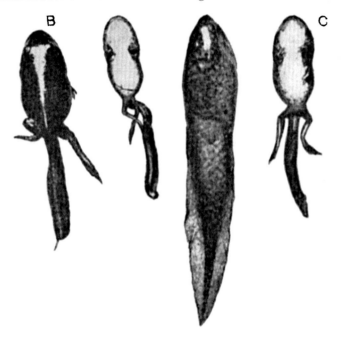

Fig. 33. Thyroid-fed tadpoles and control. Larvae of the bull frog normally retain the tadpole form for two or even three years before metamorphosis takes place. The changes produced in the three "half-and-half" individuals shown were the result of twenty-six days of thyroid feeding (W. W. Swingle, *Journal of Experimental Zoology*, 1918).

The specific effect of thyroid feeding or injection of thyroid extracts in inducing metamorphosis is equally well seen in Urodela (*Amblystoma*, salamander, triton). The most interesting case is that of the Mexican variety of the salamander, *Amblystoma tigrinum*. This local form is neotenous, and when the larvae are

kept in aquaria they never undergo metamorphosis spontaneousıy.
"Axolotls," as they are customarily called, may live for ten years,
attaining a length of over a foot, breeding in captivity from generation to generation in the larval state. It has been repeatedly shown

Fig. 34. Two transformed axolotls (*Amblystoma tigrinum*): on the right, six
weeks after a single meal of thyroid gland from a Dexter monstrous calf five
months from fecundation; on the left, six weeks after one meal of thyroid from
a bulldog foetus nearing full time.

since Babak's original observations (Laufberger (1913), Jensen
(1917), Kaufmann (1918), Huxley and Hogben (1912)) that a single
meal of mammalian thyroid suffices to induce metamorphosis. At
a temperature of 20° C. the first shedding of the larval skin
takes place about 9 to 13 days later. Metamorphosis is practically

complete within four to six weeks. By determining the minimal dose per unit body weight for axolotls of approximately the same age and kept in standardised conditions this test can be used as a means of assay (Huxley). It has also been used (Hogben and Crew) to demonstrate the presence of active material in the foetal thyroid of sheep and oxen. As with anuran tadpoles metamorphosis is accompanied by a considerable loss in body weight.

If the thyroid attains its activity during the destructive decrease in size during metamorphosis it might be expected that the thyroid during metamorphosis would show a characteristic type of growth not corresponding with that of the animal as a whole. Uhlenhuth has determined in salamander larvae the relative weights of the post-branchial organ and thyroid gland at different stages. Plotting the weight of the organs in absolute values or relative to bodily size as $\sqrt[3]{\dfrac{\text{(weight of organ)}}{\text{body length}}}$ against the age of the individuals shows that the growth of the post-branchial as also the growth of the body as a whole is interrupted at the time of metamorphosis; but the thyroid does not share in the cessation of growth involving other organs; hence, the thyroid quotient rises suddenly at the time when the larval skin is shed.

The operative removal of the thyroid in amphibian embryos was first carried out successfully by Bennet Allen (1917). The method has already been indicated (chapter 1). The epithelial bodies (parathyroids) were not affected. Nor was there any evidence that the results were secondarily caused by failure of other glands of internal secretion to develop to full functional capacity. Allen's experiments were made on anuran tadpoles (*Rana pipiens*). No peculiarities attributable to the operation were detected till about three months after the thyroid anlage had been destroyed in 349 individuals. The experimental larvae resembled the controls in colour, vigour and size. At the stage, however, when the hind limb rudiments begin to grow rapidly in normal controls, a difference between the two series at once became evident. While growth continued and the gonads proceeded to develop unimpaired the somatic organisation underwent no further change in the operated animals. They failed altogether to undergo metamorphosis. The brain and alimentary tract retained, along with the general external configuration, the larval characteristics. The tail and gill clefts persisted. The hind limbs failed to grow and no fore limbs broke through. According to Terry, who described the skeletal organisation of Allen's material, the vertebrae developed no spinous processes and remained unossified, though extensive calcification occurred. In the absence of the thyroid deposition of lime salts went on, but the stimulus to the proliferation of odontoblasts was lacking. The thymus persisted in its larval condition. The pitui-

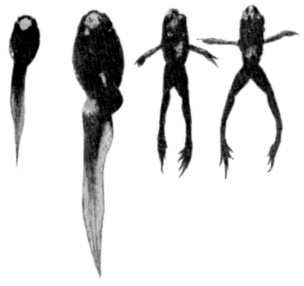

Fig. 35. "Cretin" and normal tadpoles. On the left are two thyroidless tadpoles, showing extremes in size. On the right are two normal frogs from the same "brood," and of the same age as the tadpoles (Bennett M. Allen, *Journal of Experimental Zoology*).

tary was normally developed except that it appeared to be somewhat enlarged. In short, the thyroidless tadpole failed to undergo metamorphosis.

The failure of thyroidless anuran tadpoles to complete normal somatic development was confirmed by other workers. In particular the experiments of E. R. and M. M. Hoskins (1918), which were initiated independently of those of Allen, may be here mentioned. Their operations were performed on tadpoles of another species, *Rana sylvatica*. Among several points to which they direct attention one will be left for discussion later; they point out that the lungs develop normally in the persistent larval condition of thyroidless anuran as in neotenous Urodeles; but while the gonads go on developing, these individuals cannot be properly described as neotenous inasmuch as the females do not discharge fully formed ova nor develop oviducts. In all essential respects their observations confirm those of Bennet Allen.

Though the thyroidectomised individual normally remains in the larval stage, Allen's experiments showed that it will respond to thyroid feeding even many months after the time at which it should have undergone metamorphosis; and develop into a fully formed adult. Qualitatively then, the results of thyroidectomy and treatment with thyroid substance are completely in harmony as regards metamorphosis in Anura and Urodeles. It still remains to ask whether they are quantitatively, i.e. whether the amount of active substance required to influence metamorphosis in thyroidectomised larvae is of the order of magnitude present in the larval gland. The transplantation experiments of Swingle (1921-3) and others clearly show that such is indeed the case. E. R. and M. M. Hoskins (1922), in order to test the relation of the thyroid to different stages of metamorphosis, transplanted the thyroid anlage from its normal situation to the tails of a number of larvae, whence the gland could be readily removed at appropriate intervals after the inception of metamorphic changes. After the legs were well developed the subsequent removal of the thyroid did not prevent the completion of metamorphosis. From this they conclude that the thyroid autacoid is only essential to the initial stages of metamorphosis. This conclusion must, however, be accepted with reservation since Kendall has shown how remarkably persistent may be the effect of a single dose of thyroxine when injected into the dog. In any case, in Urodele larvae (axolotls), somatic differentiation may be checked and fixed at practically any stage in the sequence of metamorphic changes. It is now impossible to doubt that thyroid activity plays an important part in initiating the histogenic changes which occur at metamorphosis. There remain for discussion the biochemical and bionomic aspects of this relationship.

The points which are of biochemical interest are most conveniently discussed with reference to a question which is still a fertile field for further research, namely, the precise relation of these phenomena to the metabolic consequences of thyroidectomy and thyroid medication in the Mammal. It has been seen that these are respectively associated with a diminution and increase in respiratory exchange in the Mammal; and that Kendall has isolated from the thyroid an iodine derivative which, qualitatively at least, has all the properties of the active substance. The respiratory exchange accompanying metamorphosis induced by thyroid feeding has been studied by Helff (1924). Helff determined the oxygen consumption of tadpoles fed on thyroid gland and kept in a solution containing diiodotyrosine with that of normal tadpoles, using Winkler's titration method. In the accompanying table is shown the increase in O_2 consumed in c.c. per gram body weight per hour. The numbers 1–8 at the head of the columns refer to arbitrary stages between 1, the normal tadpole, and 8, the tadpole in which the left fore leg had made its appearance for at least a day.

Table of oxygen consumption (Helff) in tadpoles
fed on thyroid during metamorphosis.

No. of tadpoles tested	Oxygen consumption in c.c. per grm. body wt. per hour								Increase
	1	2	3	4	5	6	7	8	
32	·157	·233							·076
19	·160		·254						·094
33	·160			·282					·122
36	·178				·286				·108
16	·171					·264			·093
27	·174						·298		·124
10	·192							·344	·152

Helff's observations show that over the complete period of metamorphosis induced by thyroid feeding there is an increase in oxygen consumption per unit body weight of 79 per cent. Furthermore there is an absolute mean increase in oxygen consumption per individual in spite of decreasing body weight. Huxley (1925) has more recently shown that the injection of thyroid extract into tadpoles produces an immediate increase in respiratory exchange measured directly by Barcroft's apparatus. These observations, taken in conjunction with those of Gayda (1922), who was unable to detect a significant change in the heat production of frogs after thyroidectomy, thyroid administration or thyroid implantation, suggest the conclusion that the histogenic incidents of metamorphosis are a consequence of the increased oxidative activity in the tissues, thus linking up the work on amphibian development with experiments on Mammals already discussed.

But this inference has not been accepted by some other workers for reasons which suggest further lines of biochemical enquiry. With the exception of Romeis (1922), all who have studied the problem are agreed that the active substance in the amphibian thyroid is an iodine compound. Morse (1914) first recorded acceleration of metamorphosis in tadpoles of *Rana pipiens* by feeding with iodised amino-acids derived by acid hydrolysis of thyroid extract. Lenhart (1915) observed that the rapidity of metamorphosis in thyroid-fed tadpoles was a function of the iodine content of the gland. Swingle (1919), whose results have been confirmed by others, found that when iodine or its inorganic compounds was administered in flour paste as food, metamorphosis was very decidedly stimulated in larvae of *Bufo lentiginosus* and *Rana pipiens*. Later it was shown that to achieve this result it was only necessary to dissolve the substance (e.g. KI) in the water in which the tadpoles were kept. Bromine compounds do not exert this action. In normal tadpoles so treated the thyroids become enlarged and the colloid masses are greatly swollen. So far the facts recorded are quite compatible with the conclusion that amphibian metamorphosis depends on the specific action of thyroxine on the oxidative processes of the body. An entirely new complexion was, however, given to the problem by Swingle's discovery that the same treatment induces metamorphosis in thyroidless larvae which would not otherwise transform into frogs at all. Swingle also showed that a certain minimum of iodine in the water or food was necessary in order that metamorphosis should occur in normal tadpoles, the minimum for thyroidless larvae being significantly higher. On the basis of this evidence he has maintained the conclusion that the essential condition of metamorphosis is not the exhibition of thyroxin specifically but the presence in the circulation of iodine.

This view harmonises with that expressed by Kendall (1919). In the Mammal thyroxine affects the basal metabolic rate in a manner quantitatively related to the dose. Injection of a derivative of thyroxine, in which what Kendall regards as the hydrogen of the imino group is replaced, has no effect on the basal metabolism. On the other hand, Kendall found that both thyroxine and his inert imino-derivative are physiologically active when tested for their effect on amphibian metamorphosis. From this he concludes that thyroxine has two modes of action, one more specific upon basal metabolism in virtue of an imino group; and one less specific on amphibian development in virtue of the fact that it is an iodine compound. This coincides closely with Swingle's standpoint. However, the more recent work of Harington (1926) shows that the question requires reinvestigation.

Uhlenhuth (1922), on the basis of experiments on Urodeles, has criticised this view vigorously. Though there is not complete unanimity in this matter, the balance of testimony is against the view that administration of *inorganic* iodine to Urodele larvae promotes metamorphic changes. On the other hand, there is no doubt from the work of Jensen (1917–21) that various organic iodine compounds have this effect and Swingle (1923) records metamorphosis by injection of diiodotyrosine into thyroidectomised axolotls. To the writer it seems that there is not enough evidence based on direct measurement of basal metabolism to settle the precise relation of the phenomena under discussion with those described in the previous chapter. What is pre-eminently needed is information of the variations of the oxygen consumption associated with metamorphosis in all experiments of this kind. If the injection of Kendall's imino-derivative is followed by metamorphosis without the characteristic rise in metabolic rate associated with injection of thyroxine, or if thyroidless tadpoles undergo metamorphosis when fed with inorganic iodine compounds without increase of basal metabolic rate, Kendall's conclusion is justified. At present we cannot say. But there are no known facts that are incompatible with the view that thyroxine is the normal form in which iodine is stored in the animal body, the power to do so being specially developed by the thyroid, which thus acts as the storage organ of the iodised active substance. As Uhlenhuth rightly points out, Swingle's experiments, important as they are, throw no light on the form in which the iodine acted in the tissues, since the fate of the substance administered in the body is quite obscure. Again it may well be that the ease with which other tissues can transform iodine compounds into thyroxine varies, being greater in Amphibia than in Mammals and greater in Anura than in Urodeles. On this view the primary function of the thyroid is not so much that of specifically manufacturing as of storing thyroxin. More experiments on the lines of those recorded by Helff and by Huxley are needed. In particular it would be of very great interest to know whether any differences exist between the effect of thyroxin or iodine on the oxygen consumption of isolated tissues. The disappearance of the external gills in the anuran tadpoles and the appearance of fully developed limbs in the Urodele larvae before the functional activity of the thyroid begins, the fact that metamorphosis involves in Anura but not in the Urodeles the destruction of the tail, the persistence of the tail fin in some Urodeles but not in others, the apparent indifference (Gayda) of the adult tissues to the action of the thyroxine—all these phenomena suggest lines of investigation which might profitably be studied by Barcroft's method on the lines suggested by Huxley's experiments. It is worth mentioning in this connexion that Bennet Allen calls attention to a very decided difference in the extent to which limb development may proceed in thyroidless tadpoles of *Bufo lentiginosus* and *Rana pipiens*. Quite apart from the practical interest of deciding whether amphibian development is an indicator of metabolic rate, the response of tissues to different levels of metabolism has a far-reaching significance for the general physiology of development, inasmuch as it falls into line with the conception of spatial organogeny introduced by Child and now placed on a satisfactory basis of direct measurement with modern methods of gas analysis by the recent work of Creswell Shearer (1924). From these considerations several points which are of bionomic interest arise.

Among the Amphibia varying degrees of neoteny or fixation of the larval phase are met with, ranging from the normal prolongation of the aquatic form for several seasons until the gonads are mature (*Rana clamata* and *R. catesbiana*) to the more extreme cases illustrated by *Amblystoma tigrinum*. Within the latter species two grades are seen in the Mexican and Colorado races. The latter may breed in the larval condition; but they undergo spontaneous metamorphosis in aquaria especially if subjected to external disturbance. The Mexican variety as stated breeds from generation to generation in the larval condition (which was at one time regarded as a true perennibranchiate under the generic title *Siredon*); and it does not undergo metamorphosis spontaneously. Some writers have stated that it will do so if kept in very shallow water (Marie de Chauvin, Boulenger) over a long period. This is certainly not true of all individuals (Hogben); and it may be that the specimens which did so in Boulenger's experiments were of the Colorado race. De Filippi found that sexually mature specimens of *Triton* larvae occur constantly in a small lake in the Italian Alps. A discussion of this problem has been deferred, because it links up with the relation of another "ductless gland" to metamorphosis.

The case of the axolotl raises the possibility that the perennibranchiate genera (Proteidae)—*Siren, Proteus, Necturus* and *Typhlomolge* may be in reality sexually mature larval forms of which the adult stage has disappeared through failure of the mechanism on which metamorphosis depends. In the case of *Proteus*, Jensen found that prolonged thyroid feeding had no effect. The same is true (Huxley and Hogben) of *Necturus*. Moreover, as Swingle showed by grafting the thyroids of *Necturus* into bullfrog larvae, the thyroid of this perennibranchiate does contain the active substance. If Gayda's observations are correct the power of the tissues to respond to the iodine-active compound may be restricted to a definite stage in the life cycle. And it is possible that administration at an earlier stage would prove efficacious. At present there is no experimental evidence that the Proteidae are neotenous forms of which the adult stage has been completely suppressed. The case of those Amphibia which have a prolonged larval existence may be met with one of two explanations. In some cases, e.g. that of *Triton* larvae in mountain lakes, it may be due to deficiency in available iodine in the external environment. But such an explanation could only account for a very small proportion of the cases. Swingle (1922–3) has shown by grafting the glands of Anura with prolonged larval life and those of axolotls into tadpoles, that the thyroid of these genera contain the active material long before metamorphosis usually occurs. He infers the existence of a second factor in metamorphosis which controls the growth and secretory discharge of the thyroid autacoid. Though the evidence is qualitative rather than quantitative this view put forward on other grounds by Uhlenhuth is reinforced by considerations to which we shall now turn.

§ III

Hardly less striking than the progress which has been made in relation to thyroid function are the experiments which have simultaneously been carried out on the part played by the pituitary gland in the timing mechanism in amphibian metamorphosis.

Shortly after the publication of Gudernatsch's investigations on thyroid feeding Adler (1914) made the first attempt to remove the pituitary in anuran (*R. temporaria*) tadpoles. Out of 1200 specimens in which the hypophysis was destroyed by cautery at a late stage (20 mm. larvae) three individuals survived, and in none of these did the hind limbs develop beyond a small bud: transformation did not take place; they remained neotenous tadpoles. Two years later P. E. Smith (1916) and Bennet Allen (1916) simultaneously and independently developed the technique already described (chapter I) for removing the anlage at an earlier stage with a comparatively small death-rate, and a high percentage of complete successes. Apart from the pigmentary characteristic already referred to (chapter III), the hypophysectomised larvae showed the following peculiarities: (i) they all failed, like Adler's three specimens, to undergo metamorphosis; (ii) their thyroids were definitely under-developed, being about one-third of the normal size and containing little "colloid" material. An intimate connexion between these two phenomena was suggested by later work of Bennet Allen (1919) that showed the possibility of inducing metamorphosis in pituitary-less tadpoles by thyroid administration. Furthermore, Bennet Allen showed that whereas tadpoles endowed with both these glands can complete their metamorphosis with the intake of very minute quantities of iodine in their food or medium, administration of large quantities of inorganic iodine will also induce metamorphosis in tadpoles deprived of the thyroid, the pituitary or both glands.

The fact that pituitary-less tadpoles do not complete their development in normal conditions may be regarded as established

beyond reasonable doubt (cf. Fig. 16, chapter III). The
results referred to have received independent confirmation by
other workers. The nature of the relationship is not so fully
elucidated. Reference has already been made to experiments
in which thyroids from neotenous larvae of *Rana clamata* were
transplanted by Swingle into younger individuals promptly
inducing metamorphosis in the latter. This suggests that the
thyroids of neotenous larvae are active; but that the stimulus
for the release of their active product is wanting. Swingle and
Bennet Allen have independently investigated the effect of implantation of the different lobes of the pituitary (anterior, tuberalis,
intermedia and nervosa) into anuran tadpoles; and have shown that
individuals into which the pars anterior has been grafted undergo
precocious transformation when the graft survives. Furthermore,

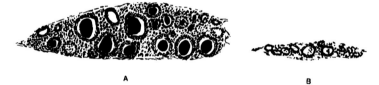

A   B

Fig. 36. A. A sagittal section through a lobe of the thyroid of a 38 mm.
check. × 100. B. A sagittal section through a lobe of the thyroid of a 37 mm.
albino. × 100.

in these cases there is a greater development of the thyroid. From
this it appears that the anterior lobe is that part of the hypophysis
concerned with metamorphic changes; and that it operates by
stimulating the activity of the thyroid apparatus. This conclusion
is reinforced by experiments of P. E. and I. P. Smith (1922), which
at the same time indicate the possibility of obtaining an active
extract of anterior lobe substance. Feeding experiments both with
anuran tadpoles and with Urodele larvae have proved quite in-
efficacious: on the other hand, injections of saline extracts of bovine
anterior lobes not only accelerated development in normal tadpoles
but induced metamorphosis in hypophysectomised individuals.
However, the injection of the anterior lobe extracts does not induce
metamorphosis in the thyroidectomised anuran tadpole.

Metamorphosis in the neotenous Mexican axolotl is not induced
by feeding for several months on anterior lobes of ox glands.

Hogben (1922) stated that injections of fresh anterior lobe extract did so, and experiments with commercial extracts on thyroidectomised axolotls also gave positive results. Experiments with commercial extracts are, however, to be regarded as having doubtful value. Smith found that retardation of metamorphosis accompanied pituitary injection in the Colorado axolotl. But Spaul (1925), who has more recently investigated the question, interprets the retardation as the result of presence of posterior lobe substances. The failure of Smith to induce metamorphosis in the Urodele by anterior lobe injection he attributes to the necessity of extracting in acid medium from absolutely fresh material, if a highly active preparation is required. Spaul's conclusions are in line with unpublished experiments of the writer.

If the anterior lobe of the pituitary influences metamorphosis by stimulating the development and discharge of the thyroid it would seem that there is considerable presumption from the work on amphibian metamorphosis in favour of the assumption that the active constituent of the thyroid is normally liberated into the circulation. The result of most general interest for the physiology of development is that the influence of the pars anterior on the development of the thyroid, and of the thyroid on somatic differentiation represents a considerable step forward in the attempt to envisage a mechanism by which the orderly sequence of developmental processes is maintained[1].

In conclusion it is permissible to ask whether these investigations throw any light on the significance of the pars anterior in the economy of the Mammal. Experimental extirpation of the pituitary in the Mammal has not yielded very convincing data. Nor with the exception perhaps of the recent observations of Evans and Long (1922) has the administration of anterior lobe extracts to Mammals. There are, however, clinical indications that the pars anterior is related to growth processes, especially skeletal development. In the case of Amphibia the evidence is more definite. Hoskins and Hoskins (1918) first called attention to the fact that thyroidectomy in anuran tadpoles is associated with quite definite

---

[1] An important communication by Helff (*Journ. Exper. Zool.* **45**, 1926) dealing with the sequence of metamorphic changes has appeared since the above was written.

acceleration of growth, a fact that they correlated with the increased development of the pituitary after thyroid extirpation. The

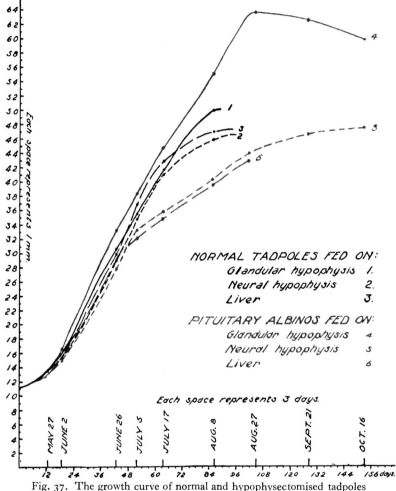

NORMAL TADPOLES FED ON:
Glandular hypophysis   1.
Neural hypophysis   2.
Liver   3.

PITUITARY ALBINOS FED ON:
Glandular hypophysis   4
Neural hypophysis   5
Liver   6

Each space represents 3 days.

Fig. 37. The growth curve of normal and hypophysectomised tadpoles
with the various glandular diets.

ablation of the pituitary on the other hand was shown by Smith
(1921) to be associated with very definite retardation of growth
though the persistence of the larval phase enables the albino larvae
to attain dimensions which are greater than those of normal

specimens before the onset of metamorphosis. Feeding with anterior lobe substance restores the normal growth rate. This is also true of axolotls (Huxley and Hogben, Uhlenhuth). Smith found that the growth-promoting constituent is, unlike Brailsford Robertson's "tethelin," insoluble in boiling alcohol (Fig. 37). More recently Smith and Smith (1923) have brought forward evidence to show that the growth-maintaining substance is not identical with that which promotes metamorphosis, and further that the two components are localised in different areas of the bovine anterior lobe.

The separation and identification of the active substance. or substances present in the anterior lobe, and their precise mode of action are problems which await further enquiry and problems for the solution of which the methods would appear to be available.

REFERENCES

ADLER. *Arch. Entwicklungsmech.* 39, 1914.

BABAK. *Centralbl. Physiol.* 10, 1913.

ALLEN, BENNET. *Biol. Bull.* 32, 1917; 36, 1919. *Journ. Morph.* 32, 1919. *Journ. Exp. Zool.* 24, 1918; 30, 1920.

EVANS and LONG. *Anat. Rec.* 23, 1922.

GAYDA. *Arch. di Scien. Biol.* 3, 1922.

GUDERNATSCH. *Arch. Entwicklungsmech.* 35, 1912.

HELFF. *Proc. Soc. Exp. Biol. Med.* 21, 1923.

HOGBEN. *Proc. Roy. Soc.* B, 94, 1923.

HOSKINS and HOSKINS. *Journ. Exp. Zool.* 29, 1919. *Anat. Rec.* 23, 1922.

HUXLEY. *Journ. Hered.* 14, 1923.

JENSEN. *Oversigt. Klg. Dansk. Vidensk. Selsk. Forhandl.* 1916.

KAUFMANN. *Bull. Acad. Sci. Cracow*, B, 32, 1918.

KENDALL. *Am. Journ. Physiol.* 49, 1919 (p. 136).

ROMEIS. *Arch. Entwicklungsmech.* 50, 1922.

SMITH. *Anat. Med.* 11, 1921. *Endocrinology,* 5, 1921; 7, 1923. *Journ. Med. Res.* 43, 1922. *Anat. Rec.* 23, 1922; 25, 1923. *Proc. Soc. Exp. Biol. Med.* 20, 1922.

SPAUL. *Brit. Journ. Exp. Biol.* 2, 1925.

SWINGLE. *Journ. Exp. Zool.* 24, 1918; 27, 1919; 34, 1921; 36, 1922; 37, 1923. *Journ. Gen. Physiol.* 1–2, 1918–19.

SWINGLE, HELFF and ZWEMER. *Am. Journ. Physiol.* 70, 1924.

UHLENHUTH. *Biol. Bull.* 42, 1922. *Am. Nat.* 55, 1921. *Journ. Gen. Physiol.* 1, 1918; 3, 1921; 6, 1924.

INDEX

Printed in the United States
By Bookmasters